JOHN J. EARLEY, *Architectural Sculptor* MURPHY & OLMSTED, *Architects*

SHRINE OF THE SACRED HEART, WASHINGTON, D. C.

THIS CONCRETE, MADE WITH ATLAS PORTLAND CEMENT AND EXPOSED COLORED AGGREGATES, CREATED DECORATIVE EFFECTS COMPARABLE TO THOSE POSSIBLE HERETOFORE ONLY WITH CUT AND POLISHED MARBLES, CERAMIC AND MARBLE MOSAICS—PROVIDING MAXIMUM BEAUTY WITH GREAT ECONOMY.

The Ferro-Concrete Style

Reinforced Concrete *in* Modern Architecture

With Four Hundred Illustrations of European
and American Ferro-Concrete Design

By
FRANCIS S. ONDERDONK, Jr.
Doctor of Engineering
Imperial and Royal Technical University, Vienna
Instructor in the College of Architecture
University of Michigan

*"Each material has its own
message and, to the creative
artist, its own song."*
—Frank Lloyd Wright

HENNESSEY + INGALLS
SANTA MONICA :: :: :: :: 1998

Originally published in 1928 by the Architectural Book Publishing Co., Inc.
Copyright © 1998 by Hennessey + Ingalls

All rights reserved. No part of this work covered by the copyright hereon may be reproduced or used in any form or by any means, graphic, electronic, or mechanical, including photocopying, recording, taping, or information storage and retrieval systems, without written permission of the publisher.

Manufactured in the United States of America

ISBN: 0-940512-09-2

Hennessey + Ingalls
1254 3rd Street Promenade
Santa Monica, CA 90401

Library of Congress Cataloging-in-Publication Data

Onderdonk, Francis S., 1893-
    The ferro-concrete style : reinforced concrete in modern architecture : with four hundred illustrations of European and American ferro-concrete design / by Francis S. Onderdonk, Jr.
        p.   cm.
    Originally published: New York : Architectural Book Pub. Co., c 1928.
    Includes bibliographical references and indexes.
    ISBN 0-940512-09-2
    1. Architecture. 2. Reinforced concrete construction. I. Title.
NA4125.O6   1998
721'.044'54—dc21   98-9649

CONTENTS

	Page
Index to Illustrations	v-vi
Preface	1
Introduction	3

Chapter I—THE POSSIBILITIES OF REINFORCED CONCRETE ... 17
    Quantitative Variations ... 17
    Qualitative Variations ... 21
    Dimensional Possibilities (Cantilever Construction, Setbacks) ... 23
    Forming Possibilities ... 27
        The "Wood-Centring Style" ... 27
        Metal Forms ... 35
        Composition Forms ... 35
        Auto-Centering: PERMANENT MOLDS ... 36
            SELF-SUPPORTING REINFORCEMENT ... 42
            SELF-CENTRING REINFORCEMENT ... 42
        Methods of Placement (Gravity System, Cement-Gun, etc.) ... 43
        "Liquid Stone" ... 49

Chapter II—SURFACE TREATMENT AND SCULPTURE ... 52
    Primary Treatments: Formmarks ... 54
                          Exposing the Aggregates ... 55
    Colored Concrete ... 61
        Colored Aggregate ... 66
        Portland Cement Stucco ... 82
        Scraffito ... 87
        Painting ... 88
             PREPARATION OF THE SURFACE ... 89
            PAINTS ... 90
            PROTECTIVE COATS ... 91
            EXAMPLES ... 91
        Spraying ... 95
        Reinforcing Veins ... 95
    Decorative Inserts: Intarsia ... 97
                      Tile ... 97
                      Glass ... 100
    Glossy Concrete: Polishing ... 101
                  Glazing (Glasin, Kerament) ... 102
                  Metallized Concrete ... 105
    Sculpture (Economic Advantages, Influence of Reinforcing) ... 106
        Cast Sculpture ... 108
        Chiselled Sculpture ... 127
        Possibilities ... 129

CONTENTS
(Continued)

	Page
Chapter III—**CONCRETE TRACERY**	133
Transitional Tracery	134
Figural Concrete Tracery (Pictorial-Silhouette-Tracery)	138
Details of Execution	150
Glass-Blocks	157
A. and G. Perret's Concrete Tracery	162
Chapter IV—**THE PARABOLIC ARCH**	186
Structural Advantages	192
The Esthetic and Symbolic Value of the Parabola	195
Examples	198
Details of Execution	199
Prof. D. Bohm's Churches	210
Chapter V—**THE FERRO-CONCRETE STYLE**	222
Structural Possibilities (Floors and Roofs; Blocks)	223
"Good" and "Bad" in Ferro-Concrete Design	232
Featuring the Monolithic Character of Concrete	239
Featuring the Reinforcement	245
The Ferro-Concrete Style	248
Subject Index	257
Index to Architects, Engineers, Sculptors, Painters and Authors	260
Bibliography	264

INDEX TO ILLUSTRATIONS

	Page
Apartment-houses—see "Houses"	
Auditoriums	62, 71, 72, 130, 150, 151, 204, 221, 253–255
Balustrades	39, 146, 147, 148
Banks	156, 234
Bathing Establishments	150, 173, 231, 232, 250, 251
Beams	35, 242
Boathouses	13
Bridges	51, 158, 192–196
Cat-Ar House	226
Ceilings	39, 49, 68, 69, 94, 98
Cement Gun	44
Chimneys	246
Churches	25–28, 40–42, 52, 56–59, 63, 124, 132, 140–144, 160, 177–185, 200–204, 207–208, 211–220, 224, 239–241, 248
Clubhouses	54, 55, 64, 155, 209, 235, 236
Colleges	162, 190, 222
Columns	35, 242
Cornices	32, 34
Decorative Inserts	93, 94, 146
Domes	59, 67
Exposition Bldgs	172, 197, 199, 205, 226, 252
Factories—see "Industrial Bldgs."	
Cemeteries	157
Garden-furniture	36, 51, 54, 96, 97, 139
Glass-blocks	173–175
Grilles—see "Tracery"	
Halls	85, 86
Hangars	198
Hotels	2, 11, 28, 29–32, 70, 78, 98–100, 228
Houses	16, 45, 84, 134, 159, 208, 210, 248, 250
Industrial Bldgs	14, 18, 52, 83, 79, 80, 158, 161, 186–189, 223, 225, 230, 246, 249
Lampmasts	37
Libraries	65
Lodges—see "Clubhouses"	
Markethalls	188, 191
Monuments	38, 92, 121, 122
Museums	152
Observatories	243, 244
Office Bldgs	149, 152, 156, 233, 246
Pavilions	37
Planetariums	18, 186
Postoffices	174, 206
Railroad Arches	37

INDEX TO ILLUSTRATIONS
(Continued)

	Page
Railroad Stations	8, 245
Reliefs	88, 90, 96, 97, 123
Residences	3, 95, 112-119, 133, 148, 153, 163-165, 171, 199, 208, 209, 226, 227, 229, 237
Restaurants	5, 6, 8, 9, 73
Roofgardens	2, 8, 9, 16, 47, 148, 155, 228, 229, 237
Schools	105, 160–168, 206
Scraffito	66
Sculpture	1, 89, 92, 120, 122, 125–129
Sections thru Bldgs	15, 17, 20, 24, 26, 28, 40, 154, 228, 234, 242, 252
Self-centering	43
Stairs	4, 24, 33, 43, 237
Steeltex	44
Stucco Textures	60
Surface Treatments	53, 84, 87–92, 171
Temples	9, 10, 91, 145, 154, 169, 170, 238
Textile-block-slab	8, 9, 160–165
Theatres	12, 19, 20, 46–48, 74–77, 101–104, 106–111, 135–138, 247
Towers	7, 15, 16, 55, 66, 146, 175, 176
Tracery	3, 16, 26, 27, 39, 40, 42, 50, 51, 57, 64, 112, 114, 115, 118, 132, 134, 135, 138, 139–153, 155, 157–185, 202, 203, 213, 227, 237, 239, 247, 248
Vaults	34, 43, 174
Villas—see "Residences"	
Warehouses	81–83
Water-towers	4

PREFACE

The purpose of this book is to trace the beginning of the Ferro-Concrete Style and to indicate the developments which are to be expected in the future.

Readers who glance thru the illustrations before reading the text should bear in mind that some of the buildings are shown as aberrations and not as examples of the Ferro-Concrete Style. A few photographs of buildings constructed of other materials have been added to illustrate esthetic contentions.

Thanks are due to Professor E. von Mecenseffy, Dr. P. H. Riepert, Dr. A. Willnow, the architectural journals, and the Atlas Portland Cement Co. for contributing photographs. The author is further indebted to Professor Raleigh J. Nelson and Professor George M. McConkey for suggestions concerning the text.

Courtesy *Concrete*
Fig. 1. WINGED FAIRY MODELED OF CONCRETE ON WIRE FRAME BY T. A. BROWER, WESTHAMPTON, N. Y.

Fig. 2-3. MARLBOROUGH-BLENHEIM HOTEL, ATLANTIC CITY, N. J.
Price & McLanahan, Architects.

Fig. 4. FERRO-CONCRETE
RESIDENCE,
DARESSALAM,
EAST AFRICA

Passages between service-wing and living quarters.
P. MOCZELANY, ARCHITECT.

INTRODUCTION

"Concrete construction is one thing—ancient—chiefly compressional. Ferro-concrete construction is another—modern—chiefly tensional.... This means we have the astounding phenomena of a new material to our hand... a new architecture invariably comes with a new constructive material."—*S. Woods Hill.*[*]

It is one of the tragic facts in history that contemporaries of great achievements generally remain ignorant of their advent. The Roman historians record nothing of Jesus. Few of Columbus' generation knew that America had been discovered. The general public, as well as the majority of architects, do not realize that we are fortunate in witnessing the birth of a new architecture,—the Ferro-Concrete Style.

Some may question whether it is justifiable to speak of a "Ferro-Concrete Style". Altho the Romans used concrete of a sort extensively they could not manufacture artificial cement and reinforcing was unknown to them. The invention of Portland Cement by Joseph Aspdin in 1824 and the insertion of steel in concrete by J. Monier, a gardener of Boulogne, France in 1849 introduced a new material,—Ferro-Concrete. The calculation of stresses, an achievement of our scientific age, enables the building of slender columns and long spans in reinforced concrete in contrast to the thick walls and massive piers into which the Romans poured their concrete.

As eight centuries ago France was the birthplace of Gothic, so now the French take a prominent role in the creation of the Reinforced Concrete Style. Yet Germany and our own Pacific Coast as well as individual architects in other countries have contributed to the new style almost as much as A. and G. Perret, the French leaders.

(*) "The Architectural Record", June 1928.

Fig. 5. STAIR IN THE POLICE HEADQUARTERS, AACHEN, GERMANY

Fig. 5. WATER-TOWER OF THE MAGGI FACTORY, SINGEN (HOHENTWIEL), GERMANY

Fig. 7. FERRO-CONCRETE RESTAURANT "DIE BASTEI" BUILT ON THE OLD FORT TOWER, COLOGNE, GERMANY
W. Riphahn, Architect.

"Concrete invites study", writes Mr. Cass Gilbert. "New forms adapted to the new material will be found, and a new architecture may result."[†] All architects who have devoted sufficient time to the consideration of the possibilities of ferro-concrete believe that it is creating a style. Professor Beresford Pite addressed the Royal Institute of British Architects recently on "the revolutionary effect of concrete on architecture," and a prominent American architect, Albert Kahn, is convinced "that we have only made a fair start and that the development of concrete, both structural and artistic, will exceed any present expectations."[*]

S. Woods Hill suggests that:

"In ferro-concrete architecture the source of all inspiration is to be found in the solution of the structural problem. With the clarifying of this solution comes the bright light of inspiration, and it is this quality of auto-inspiration which distinguishes ferro-concrete architecture from all other existing forms of architectural expression; the animating principle is in the suggestion of the structural material itself."[**]

F. L. Ackerman writes: "The introduction of a single material that will serve a wide range of uses hitherto served by a variety of materials is significant from the standpoint of architectural expression. For it constitutes no less than the foundation of a revolutionary change in design."[*†]

(†) "The Architectural Forum", Sept. 1923, p. 86.
(*) "Proceedings of the American Concrete Institute", 1924.
(**) "The Architectural Record", Feb. 1928.
(*†) "Proceedings of the Am. Conc. Inst.", 1927.

Fig. 8-9. RESTAURANT "DIE BASTEI" OVERLOOKING THE RHINE WITH COLOGNE CATHEDRAL IN THE REAR
W. Riphahn, Architect.

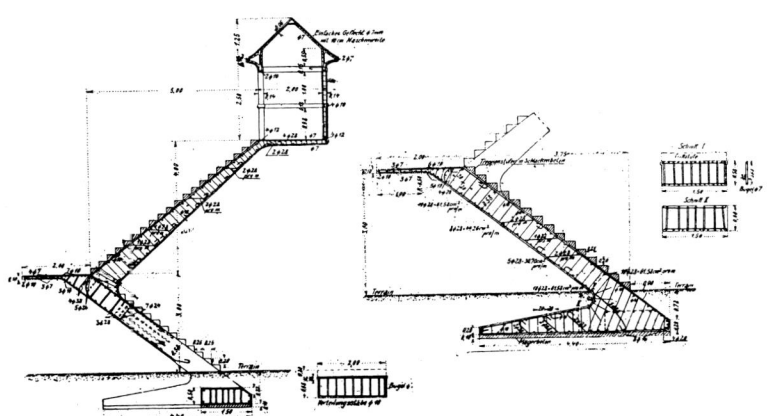

Fig. 10-11. TOWER BUILT FOR THE EXPOSITION IN KOSLIN, GERMANY
Dimensions given in the sections are in the metric system.

Fig. 12. PLATFORM SHELTER, NURNBERG, GERMANY

Fig. 13. MIDWAY DANCE GARDENS, CHICAGO
FRANK LLOYD WRIGHT, ARCHITECT.

"*Concrete is another passive or negative material depending for aesthetic life almost wholly upon the impress of human imagination. This element of pattern, however it may mechanically be made to occur, is therefore the salvation of concrete in the mechanical processes of this mechanical age . . .* " **
FRANK LLOYD WRIGHT.

John C. Austin believes that, "the adaptability and use of reinforced concrete, as a strong and economical structural and architectural medium, is rapidly increasing and gradually tending towards the development of a new architectural style."*†

David C. Allison thinks, "no structural innovation more radical has appeared before; none that at all compares in the rapidity of its growth."*

J. B. Mason speaks of a "..... growing tendency to accept and make use of concrete as an architectural and decorative medium in itself."***

Oswald C. Hering states: "The advent of reinforced concrete brings science to the aid of plastic art, insuring, at an ever lessening cost, greater stability and adaptability, together with endless opportunity for the expression of form, and gives consequent promise of an esthetic awakening of great significance......Already has the potentiality of a molded architecture arrested and stirred the pulse of both the laity and the profession....... the vast possibilities of reinforced concrete, both in its structural and esthetic phases, cannot be measured. Who can say that from the interrelated masses of aggregate,

(*†) "Proceedings, Am. Conc. Inst.", 1927.
(*)   "Proceedings, Am. Conc. Inst.", 1924.
(***) "Architecture", Sept. 1927.
(**) This and the following quotations are part of Frank Lloyd Wright's article "The Meaning of Materials—Concrete", "The Architectural Record", Aug. 1928.

Fig. 16. UNITY TEMPLE, OAK PARK, ILL.

"The first building in America to be cast complete, ornament and all, in forms—and to be "let alone" as "Architecture" after the forms were removed."

FRANK LLOYD WRIGHT, ARCHITECT.
1907

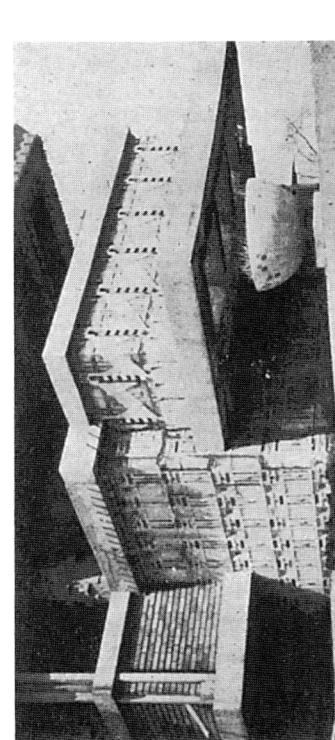

Fig. 14-15. MIDWAY GARDENS, CHICAGO, ILL.
FRANK LLOYD WRIGHT, ARCHITECT.

"Several of the illustrations show what the degraded old concrete block (a cheap imitation and abominable as material when not down-right vicious)—may become with a little sympathy and interpretation, if scientifically treated. Herein the despised thing becomes a thoroughbred and a sound mechanical means to a rare and beautiful use as an architect's medium, as the "block" becomes a mere mechanical unit in a quiet, plastic whole."

FRANK LLOYD WRIGHT.

Fig. 17. UNITY TEMPLE, OAK PARK, ILL.

". . . . Concrete may be formed in place in great size, by small means, by way of accretion—whereas stone must be got out, expensively cut, transported and lifted to place in bulk."

FRANK LLOYD WRIGHT, ARCHITECT.

cement and steel, a new and true architecture shall not be born....? It is destined to revolutionize architectural forms and inaugurate a new era of building.... and promises far greater possibilities in construction and beauty than have been recorded even in the architecture of the Greeks and the Goths.... Crude and wholly utilitarian at the beginning, concrete architecture will gradually acquire refinement and nobility.... through a frank manifestation of its own characteristics of lightness and delicacy.."††

W. L. Woollett believes we have today a traditional style: "It is this translating of old architectural forms into forms suitable to a new material which is difficult... We are obliged to follow the old forms closely in order not to get so far away from precedent that we give the impression of being illiterate or bizarre."*

In J. J. Earley's opinion, "...we may expect concrete buildings both in their structure and appearance to develop an individuality equally as marked from that of standard masonry as the latter is from frame.... When we think of the saving in the labor required to form and place a plastic mass over that required to form and place a solid mass, and of the wonderful possibilities which better knowledge will give to concrete may we not conclude that this is the economic change which will allow a fuller and greater development of structural and decorative forms?"

European architects are quite as enthusiastic. Jean Badovici, editor of "L'Architecture Vivante", introduced his essay in the 1926 volume with the following words: "Au XX. siecle, s'ouvre une ere nouvelle pour l'architecture. Le ciment impose le retour aux elements et aux verites." Andre Lurcat writes:

(††) "Concrete and Stucco Houses", pp. 69-71, 74, 104.
(*) "Concrete".

Courtesy *Concrete*
Fig. 18. HOTEL PALACIO SALVO, MONTEVIDEO, URUGUAY
It is 331 feet high and said to be the tallest reinforced concrete building in the world.
MARIO PALANTI, ARCHITECT.

"Every discovery of a new building method, or a new medium of construction, implies the abandonment of pre-existing values....In the past, architecture has always reached its highest expression when it has borrowed from no other art. Freed from the trammels of stultifying conventions and futile non-essentials alike, it is once more at liberty to develop from its own resources."

Courtesy *"Pacific Coast Architect"*

Fig. 19. HOLLYWOOD PLAYHOUSE, HOLLYWOOD
PATIO

MORGAN, WALLS AND CLEMENTS, ARCHITECTS.

Fig. 20. BOATHOUSE IN BRIONI NEAR POLA, ISTRIA, ITALY

Paul Jamot states in his book on A.-G. Perret:[†]

"Nowadays, and for all countries, ferro-concrete has its turn. An architect of today who deprives himself of concrete and of its wonderful resources acts like a business man who refuses to have a telephone or electric light installed on his premises." He concludes his book by referring to —"the original beauty of an architecture which will be the Ferro-Concrete Architecture."

Professor Mecenseffy, Munich, sees in ferro-concrete an absolutely new material which is bound to create a new style. In having this opportunity he compares our period to the twelfth and thirteenth centuries and hopes for an equally splendid development. But whereas Greek art required several centuries to mature, and Gothic one, he expects the Ferro-Concrete Style to evolve much more rapidly.

Professor Medgyaszay, Budapest, draws a similar parallel, adding, "Our problem is still more difficult: to find the artistic expression for a compound uniting materials resisting pressure and tension." Dr. P. H. Riepert believes ferro-concrete will influence the development of style because its structural system is characterized by rhythm, a trait which rules the artistic striving of our age.

Dr. A. Willnow stresses the monolithic character of reinforced concrete which is only limited by expansion joints and hinges and thus governed by static principles quite different from those of stone and wood. He concludes that:

"1. Ferro-concrete, in contrast to steel, is capable of creating an architecture; it is original concerning plan, structure, lighting and form.

2. Ferro-concrete requires an independent style because—

a) Its material has static qualities different from those of other materials, permitting other and bigger stresses.

b) It has a technique different from that of other materials."

(†) G. Vanoest, Paris 1927.

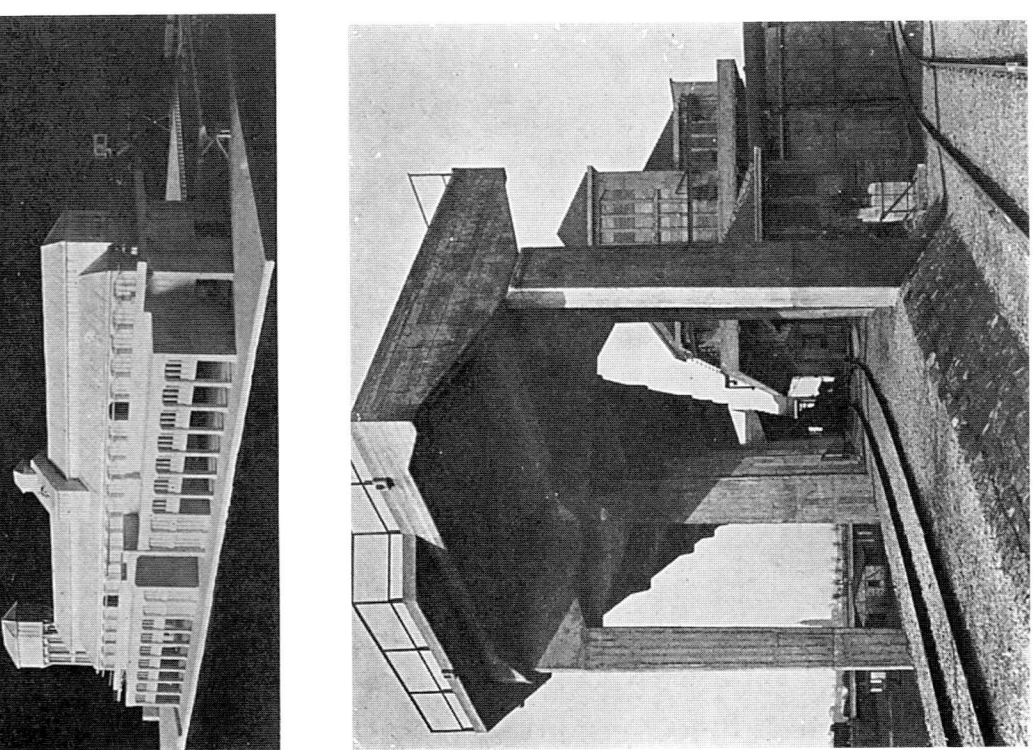

Fig. 21-23. GAS WORKS, DRESDEN, GERMANY
Poelzig, Architect.

The fact that Ferro-Concrete is considered useful but that its beauty is overlooked by many, is the best indication that it will bring forth a new style, for a new type of architecture can only develop when it expresses new structural facts, not an artist's whim. The evolution of vaulting is considered to have brought about the Gothic style. How much more should reinforced concrete with more than one hundred and eighty floor systems, and a perfect adaptability to pressure and tension produce a new style.

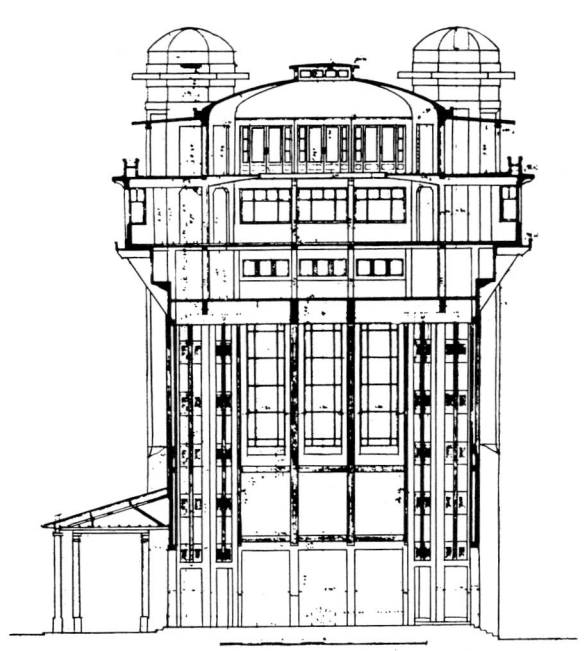

Fig. 24. "TOWER OF THE WINE-PROVINCES", 1925 PARIS EXPOSITION OF DECORATIVE ARTS
SECTION
PLUMET, ARCHITECT.

Fig. 26. HOUSE AT THE REAR OF RHEIMS CATHEDRAL, FRANCE.
Concrete frame with brick and stone filling.
PROF. L. MARGOTIN-THIEROT, ARCHITECT.

Fig. 25. "TOWER OF THE WINE-PROVINCES", PARIS EXPOSITION OF DECORATIVE ARTS
A restaurant was located on the top floor.
PLUMET, ARCHITECT.

CHAPTER I
THE POSSIBILITIES OF REINFORCED CONCRETE

Before discussing the characteristics of the Ferro-Concrete Style the possibilities of the material should be briefly described, as the style is based on the utilization of these potentialities. Many architects are prejudiced against concrete because they are not informed of the variety of techniques which have been developed for forming and decorating it. According to *The Esthetics of Potentialities,* a theory developed by the author in his dissertation 1918, a material approximates perfection in proportion to the number of possibilities it presents. From this point of view ferro-concrete is the most perfect building material, as its potentialities far outnumber those of other materials. It combines the good qualities of stone and steel, while it at the same time avoids their limitations. It can bear tension like steel, but can resist fire and rust. Its possibilities can be grouped under five headings:

QUANTITATIVE VARIATIONS

"Reinforced Concrete" according to A. Rey, the French architect, "is a term that in reality embraces an unlimited number of different modes of construction". By increasing or decreasing the sectional steel area the strength of a concrete member of fixed dimensions can be varied. For certain purposes concrete with a minimum of steel is satisfactory; in others the steel construction becomes dominant and the concrete is cast around it as a protection against fire and corrosion.

In some cases, such as for the walls of residences designed by Mr. Flagg, gravel has been replaced by rubble stones, up to thirty or forty per cent, or by brick, the interspaces being filled with ordinary concrete. At the other extreme is the elimination of all large aggregate, the steel being covered with cement mortar, one part cement and two parts sand. Grilles of great delicacy can thus be made of reinforced concrete (Fig. 1).

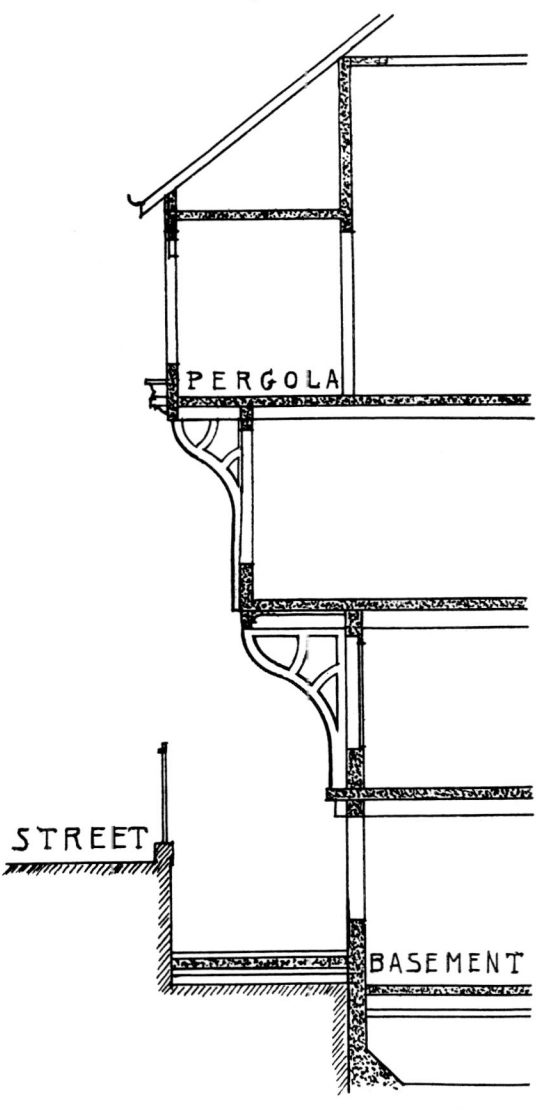

Fig. 27. SECTION THRU THE HOUSE SHOWN IN FIG. 26

Fig. 29. PLANETARIUM, GESOLEI, DUSSELDORF, GERMANY
(A planetarium has a dome top and electrical arrangements whereby any part of the sky can be reproduced, the planets and stars being in their natural positions).
ALLGEMEINE HOCHBAU GESELLSCHAFT A.-G., BUILDERS.
PROF. DR. W. KREIS, ARCHITECT.

Fig. 28. HOUSE OF MODELS, LOCOMOTIVE FACTORY HENSCHEL & SON, KASSEL, GERMANY
CURT VON BROCKE, ARCHITECT.

Fig. 30. FLORIDA THEATRE, JACKSONVILLE, FLA.
R. E. Hall & Co., Architects & Engineers.

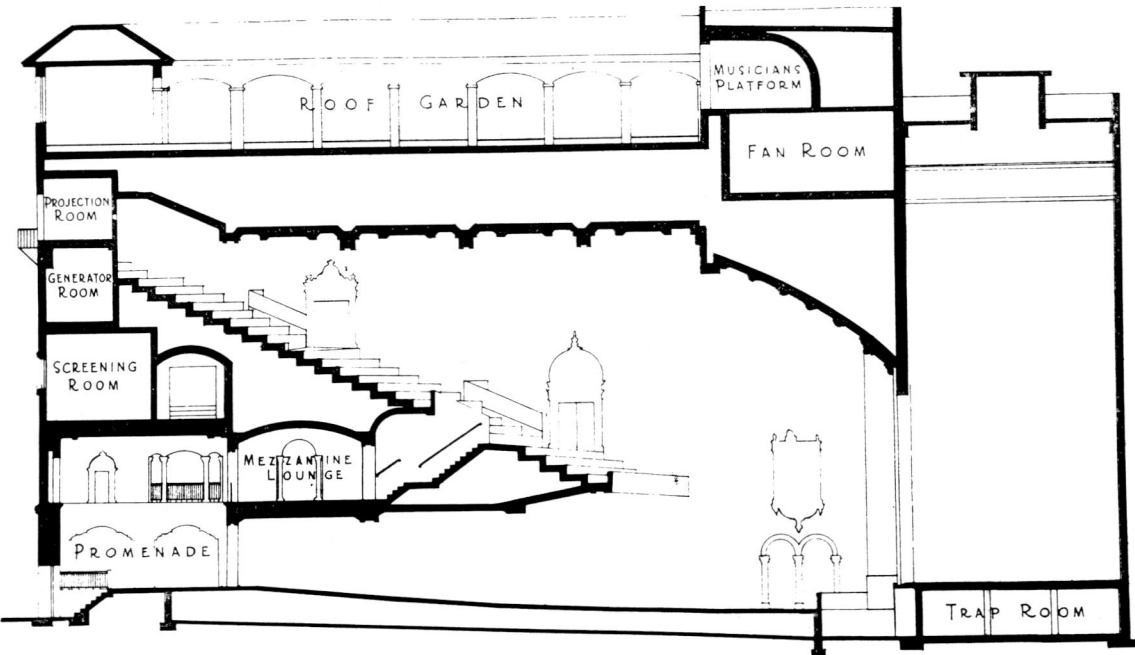

Fig. 31. FLORIDA THEATRE, JACKSONVILLE, FLA.
Longitudinal Section.

Even sectionally concrete members can vary: structures which are to hold water have a richer mixture applied at the surface whereas the core is made with a smaller percentage of cement. The center-dome of the Friedrichstrassen Passage in Berlin has piers the inner side of which consists of a 1:3 concrete with granite aggregate, whereas the remaining portion which bears less pressure is a 1:4 mixture.

The quantity of water added to the concrete determines its strength and resistance to weather, wear, and other destructive agencies as well as its permeability. The water-cement-ratio has been ascertained by Professor Duff A. Abrams after experimental studies which were carried on from 1916 till 1926. An excess of mixing water results in concrete of low strength, as it dilutes the cement paste and weakens the adhesion between the particles. The strength of concrete is now no longer considered a function of the quantity of cement; rich mixes give b e t t e r concrete only due to the lowered water-cement-ratio.

John J. Earley, the eminent architectural sculptor, does not believe in the "dogma of initial set". He maintains that one can do anything to portland cement in the first twelve hours after wetting it, and suggests adding sufficient water to concrete to produce the flow, and once it is in the forms to remove enough water to produce the best c o n s i s t e n c y for strength. This method was employed in the casting of the columns for the Nashville Parthenon:

"The outside wall of the mold was constructed with vertical staves and spacing blocks held together by encircling rods. Metal lath was attached to the inside edges of the staves. The inside wall was segmental in form and could be taken apart and removed through the top of the column. Thus a porous mold was

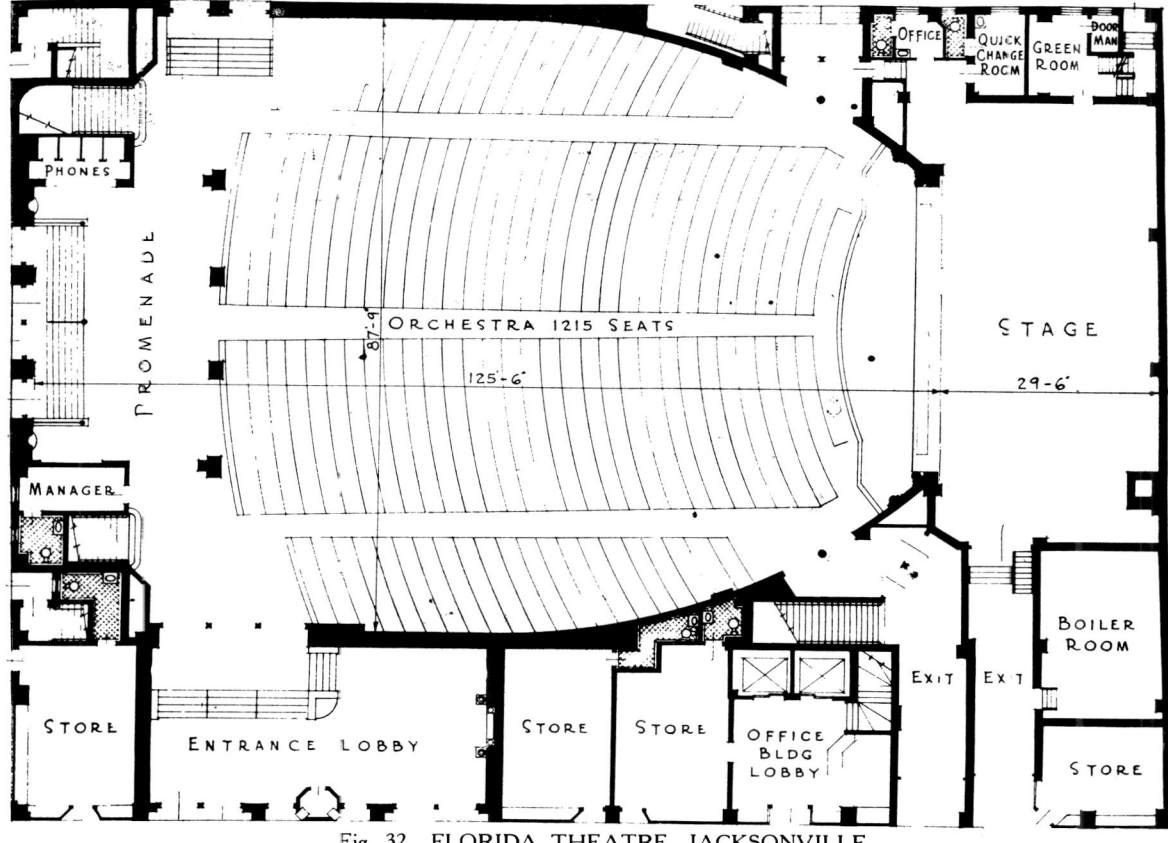

Fig. 32. FLORIDA THEATRE, JACKSONVILLE
Plan of Main Floor.

constructed which separated the excess water from the concrete after it had served to facilitate its placing, leaving the concrete in an almost ideal state to develop greatest density and strength. When the mold was removed the metal lath remained on the concrete; later it also was stripped off and with it came the skin of fine cement usually found on concrete. This left the structural cores both porous and rough, an ideal state for the attachment of the finishing concrete which was poured into molds of the finished form. Precast details of exquisite workmanship were added."[1]

QUALITATIVE VARIATIONS

In addition to the variability of the proportions in which cement, sand and gravel are added, reinforced concrete permits various qualitative differences.

Different kinds of steel serve as reinforcement: cast-iron columns (used as a core), structural steel and ordinary reinforcing bars are employed.

According to the *Handbuch fur Eisenbetonbau*[2] seventeen different classifications of concrete can be made. The same authority ascribes to portland cement nine chemical constituents and differentiates it according to seven qualities. Sand and gravel can be classified in four principal ways and with regard to fifteen special qualities. Thus ferro-concrete is a combination of materials each of which has many possibilities, so that there is reason to expect that the compound product will be truly excellent.

(1) "Proceedings of the American Concrete Institute", 1926, pp. 520, 533.
(2) Vol. II, pp. 20-52.

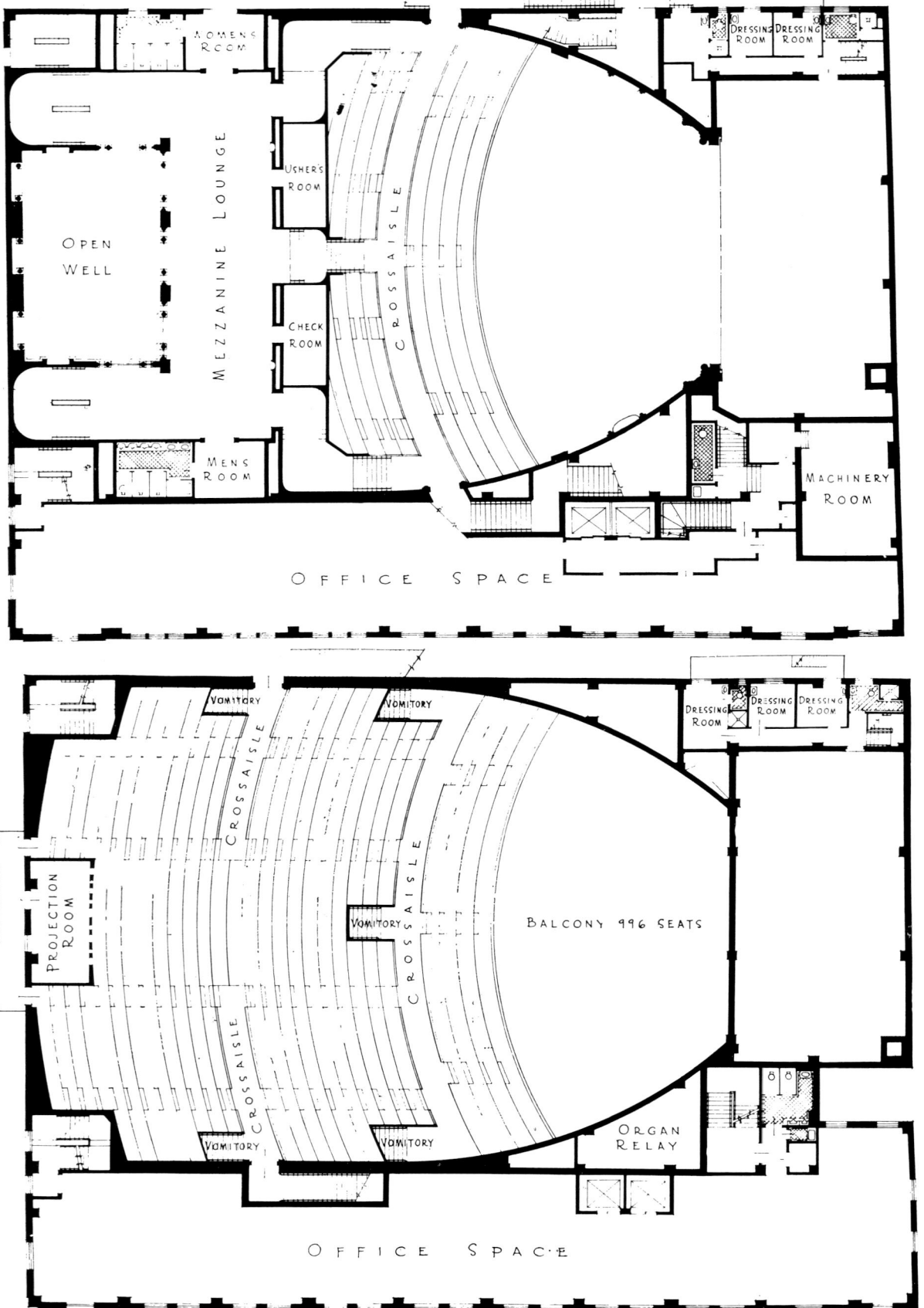

Fig. 33-34. FLORIDA THEATRE, JACKSONVILLE, FLA.
Plans of Mezzanine and Balcony.

Fig. 35. FLORIDA THEATRE, JACKSONVILLE

So as to achieve economical construction the weight of concrete must be reduced wherever possible. This can be done by using light aggregates or by artificially creating aircells in the concrete. Pumice-concrete and cinder-concrete belong to the first type, whereas the new "Bubblestone"[3] is an example of the second method. These light concretes not only reduce the dead load but also insulate much better than gravel concrete; units made of light concrete are easier handled and nails can be driven into them.

Gravel-concrete is stronger than the above variations and is most often employed. The advantages of the variability of the qualities of the aggregates are shown in the dome of the Marlborough-Blenheim Hotel (Fig. 2,3): granite aggregate was used for the ribs and beams but the connecting shell is of cinder-concrete. Other examples are concrete stairs and passages which are covered with a surface-layer of a rich mixture with specially hard sand and gravel that can resist wear.

DIMENSIONAL POSSIBILITIES

In 1878 Monier claimed for reinforced concrete that, so far as form is concerned, it has no limits. Reinforced concrete has conquered the third dimension. Former styles of architecture are characterized by an adherence to the vertical: stone was laid upon stone, brick upon brick. Projecting cornices and balconies were like daring features in an acrobat's display. Now the unity of ferro-concrete construction makes

(3) "Concrete" (Chicago), Feb. 1928, p. 30.

Fig. 36. FLORIDA THEATRE, JACKSONVILLE, FLA.
R. E. HALL & Co., ARCHITECTS AND ENGINEERS.

projection on a much larger scale a natural, commonplace matter (Fig. 4-9). A console can be so strongly reinforced that it may carry a pavilion (Fig. 10, 11). For the first time the architect is fully master of space. He has unlimited possibilities in designing the outline of his building; as the Baroque architect curved and cut in the design of his floor-plan, so the Ferro-Concrete architect will be able to abolish the straight contour in his section and elevation.

The parts of a reinforced concrete structure can be compared to muscles in contrast to the "crutches" of the older construction types. The unity of a ferro-concrete building makes wood and brick construction appear like patchwork. The projection of a balcony or cornice was limited in old construction; some Gothic oriels incite our admiration just because they approach the limit of the possible in the realm of masonry. The projection of a ferro-concrete slab can be extremely big (Fig. 12). The very existence of such possibilities will inspire the architect in his creative work. Hence Ferro-Concrete Architecture will be a veritable *Raumkunst* —"space-art".

Ferro-concrete has conquered the vertical: the designer will use it when desired but not by compulsion. The silhouette possibilities of concrete are practically unlimited.[4] From the viewpoint of *The Esthetics of Poten-*

---

(4) Cubists sometimes revel in very far over hanging slabs unsupported by consoles; esthetic doubts concerning exaggerations of this type are stated in the last chapter.

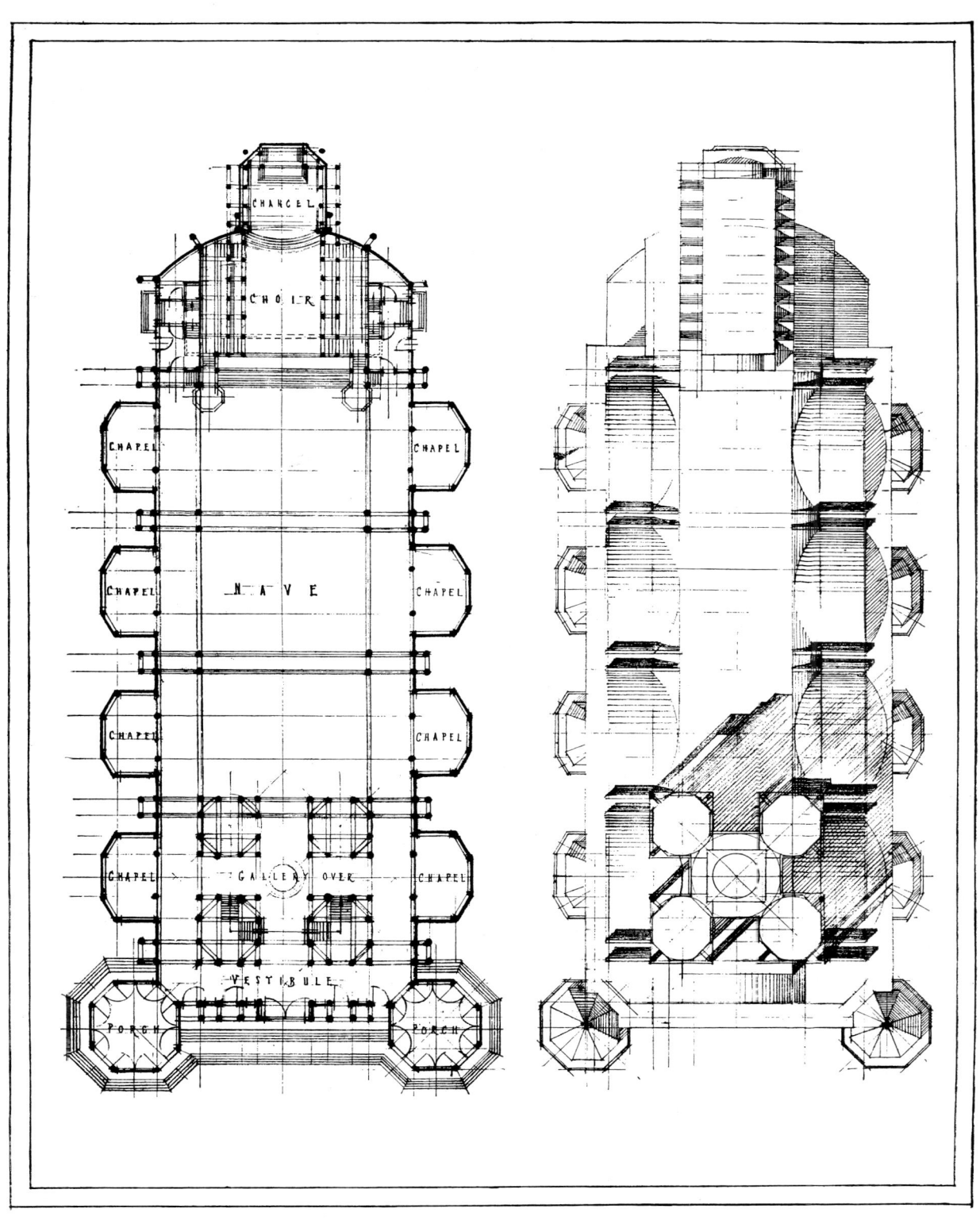

Courtesy *The Architectural Record*
Fig. 37a. DESIGN FOR A CHURCH IN REINFORCED CONCRETE
Ground and Roof Plan.
S. Woods Hill, Architect.

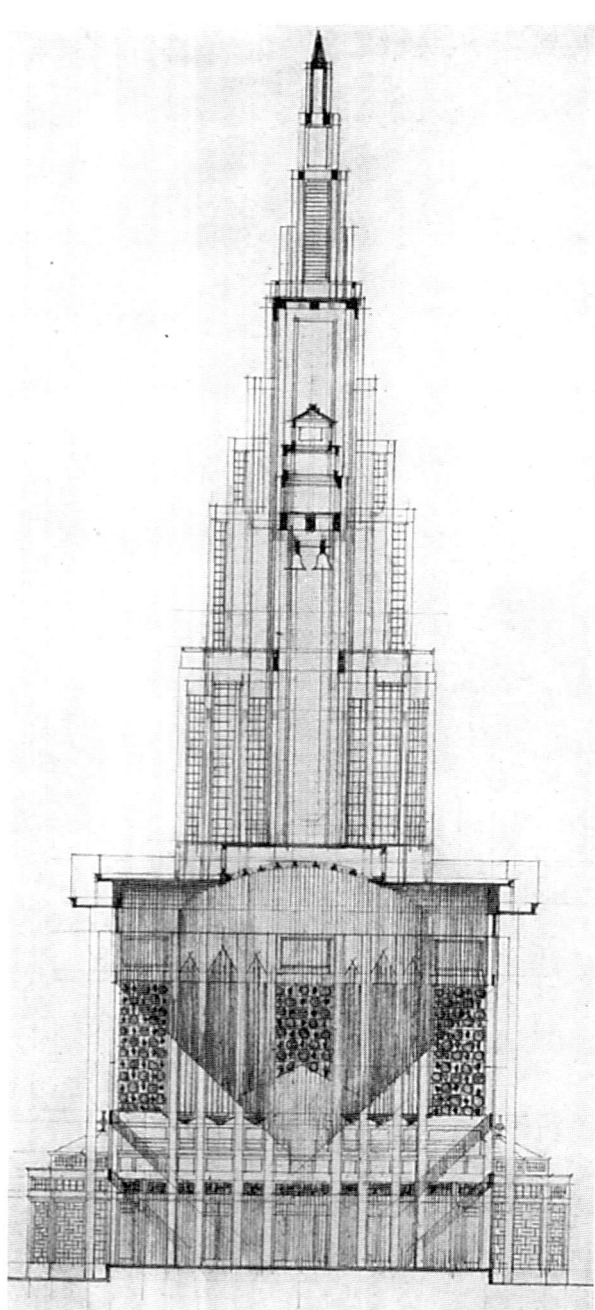

Courtesy *The Architectural Record*
Fig. 37b. A REINFORCED CONCRETE CHURCH
Transverse Section.
S. Woods Hill, Architect.

tialities the complete domination of the third dimension, the final conquest of space, signifies an advance towards perfection. This is so important a step that when the History of Architecture is written Ferro-Con-

(5) "Concrete", April 1928.

crete may be considered the beginning of a new period (Fig. 13-17).

The outrigger construction of medieval timber houses can be repeated in concrete buildings when the streets are sufficiently wide; the floorspace is increased and the floor construction economical due to the decrease of the bending moment of the center slab caused by the negative bending of the overhanging slabs (Fig. 18-28). In theatres, churches and assembly halls ferro-concrete enables wide overhanging galleries that do not necessitate obstructing columns in the auditorium below (Fig. 29-36).

Cantilevers are so suitable for reinforced concrete that vaults and domes have been constructed as a combination of two cantilevers meeting at the crown. The Los Angeles City Hall has a dome of this type and the B. G. Goodhue associates have built cantilever-vaults for several churches.[5] Mr. S. Wood Hills published an interesting design of this type for which he anticipates a saving of 30 to 40 per cent in cost on which he writes:

"By spanning the roof on girders designed as cantilevers it is possible to employ the material most efficiently and economically. These girders being "upset" and projecting above the roof bring the roof slab into action as a compression flange, which is structurally desirable. The elevations of the spandrel walls which support the segmental transverse vaults are so designed that, with proper reinforcing, they act as longitudinal beams, the effective depth increasing toward the center of the span...Several posts coupled together will give a higher radius of gyration than one and with less expense... by providing lateral bracing struts at intervals, each individual column

Courtesy *The Architectural Record*
Fig. 37c. A REINFORCED CONCRETE CHURCH
South Elevation.
S. WOODS HILL, ARCHITECT.

need not exceed an unsupported length of fifteen diameters as required by legal regulation... (French and German engineers use greater unsupported column lengths with an increased percentage of vertical and spiral reinforcement if necessary)."[6]

The self-bearing reinforced concrete wall hanging between two concrete columns acts like a narrow beam of great height and is an additional possibility of the new style.

The conquering of the third dimension is seen in the ease with which setbacks are introduced in concrete construction (Fig. 38, 39). This is likewise true of steel-frame buildings, but reinforced concrete has the advantage of being monolithic. New York's skyscrapers, due to the city ordinance, show interesting silhouettes demonstrating the value of a "three-dimensional" architecture.

FORMING POSSIBILITIES
The "Wood-Centring Style".

Many so-called characteristics of concrete architecture apply to the traditional method of wood-molds

---
(6) "The Architectural Record" March 1928

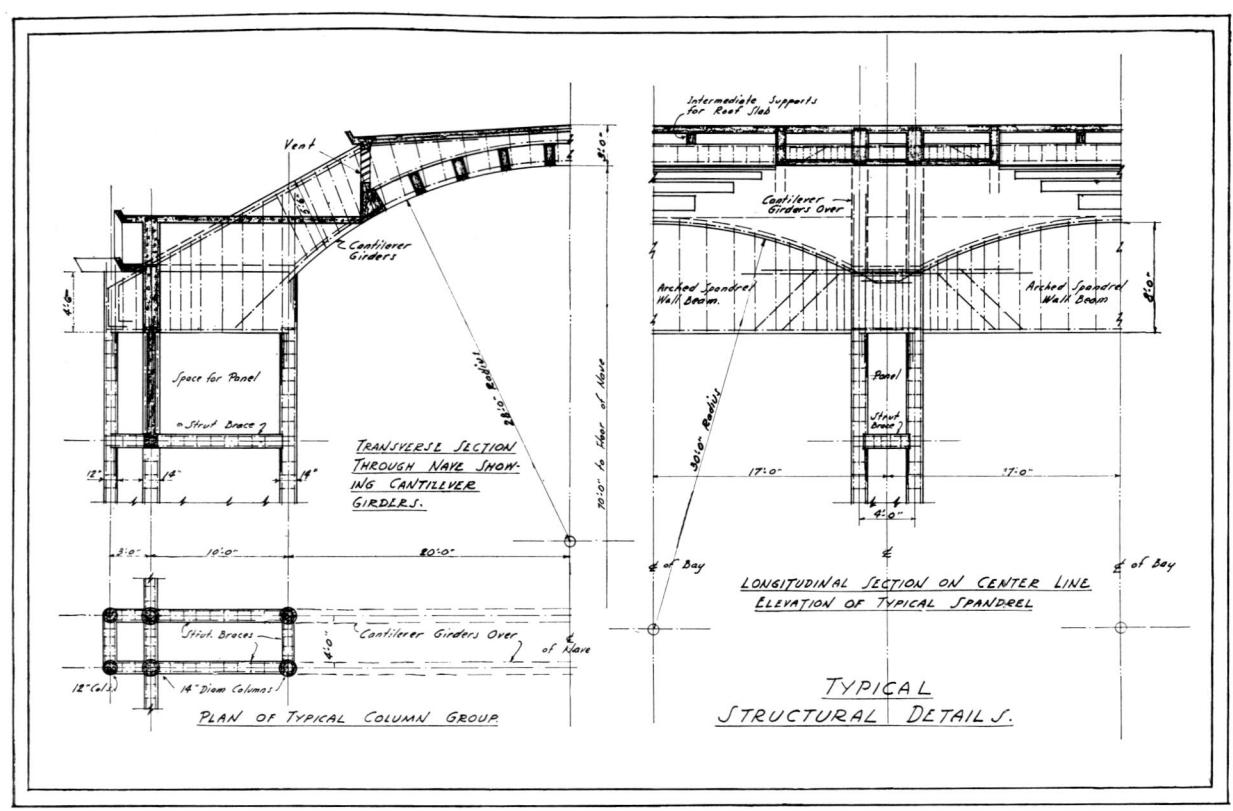

Courtesy *The Architectural Record*
Fig. 37d. A REINFORCED CONCRETE CHURCH
Structural Details.
S. Woods Hill, Architect.

and not to concrete itself. The increased use of curved metal molds, of self-centring reinforcement, and of the cement-gun will allow the innate qualities of reinforced concrete to develop, especially as regards plasticity and adaptability to every form, size and curvature. Wood-centring signifies "stammering", i. e., building every part twice: first as mold and then in concrete. Hence neither is done with that care and patience devoted to the old materials where every operation contributes something permanent. This drawback has been well expressed by Mr. I. K. Pond:

Courtesy *Concrete*
Fig. 38. FLEETWOOD HOTEL, MIAMI BEACH
Cement Stucco Finish.
Frank V. Newell, Architect.

"If the concrete product is to be perfect, the form or mold must be perfect. That the concrete may be perfect the forms must be destroyed; that which in itself was perfect must be destroyed that ultimate and permanent perfection may be achieved. A poignancy as of grief must accompany the contemplation of

Fig. 39. THE FRANCISCAN HOTEL, ALBUQUERQUE, N. M.
Portland Cement Stucco on Reinforced Concrete.
Trost & Trost, Architects.

the loss and destruction which must attend the creation and production of any structural object."[7]

Due to its plasticity before setting, concrete can adopt any conceivable form. The more a material can be affected by mechanical and chemical influences while being formed, the more possibilities it contains and hence the more perfect it is. The variety of methods for forming concrete listed below as well as the many modes of surface-treatment prove that ferro-concrete is unique.

Many architects doubt the esthetic possibilities of concrete because they identify the limitations of wooden molds with limitations of ferro-concrete, an opinion which is a fatal

(7) "Proceedings of the American Concrete Inst." 1927.

Fig. 40. THE FRANCISCAN HOTEL
Detail of Corner.

Fig. 41a-b. THE FRANCISCAN HOTEL, ALBUQUERQUE, N. M.

error. Mr. F. L. Ackerman even writes: "Our theory of concrete design resolves itself immediately into a theory of building the preparatory structure, the form.... the form is the beginning and the end of concrete design."[8]

Mr. Cass Gilbert champions the "Wood-centring Style" when he writes:

"As it is cheaper to build a lintel and a post than it is to build an arch and a pier, in proportion as it is cheaper to build in a straight line than in a curve, we follow the direction in which the finger of "economy" points and build columns and girders and flat surfaces rather than those in complex curves. It is cheaper because the "form" costs less. Therefore the "form" dominates the design, and economy has its innings."

The fallacy of this view which prevails among American architects becomes obvious when we realize that wire-mesh reinforcing obviates molds,—be it for a curved concrete garden-vase[9] or a monumental cupola and vault (Fig. 320-331). Concrete is accessible in its plastic stage whereas we cannot benefit from the fact that granite was a liquid material in its formative period. Therefore Mr. Cass Gilbert should not draw a comparison between the two nor disagree with Russell Sturgis whom he quotes as remarking: "Terra cotta should be *poured* at the top of the building and allowed to run down until it covered the structure"[10] Ferro-concrete buildings should have such a flowing character (Fig. 365-367), besides exhibiting elasticity (Fig. 301) and lightness (Fig. 259-268).

Most concrete buildings up to the present time have angular, bulky shapes because these are the easiest ones to create by wood-centring. The illustrations of this book show examples where curved shapes and slender forms of concrete have been produced by designers who rid themselves of the

(8) "Proceedings of the American Concrete Inst." 1927, p. 259.
(9) "Concrete Pottery and Garden Furniture", by R. C. Davison.
(10) "The Architectural Forum", Sept. 1923, pp. 83, 84.

Fig. 42-43. LOBBY AND BALCONY OF THE FRANCISCAN HOTEL,
ALBUQUERQUE, N. M.
Trost & Trost, Architects.

Fig. 44. THE FRANCISCAN HOTEL
Corridor.

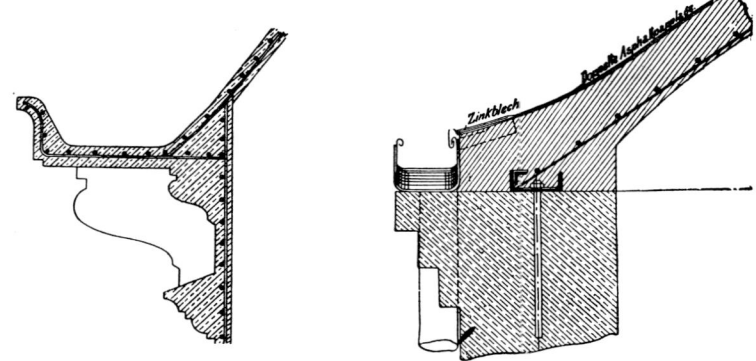

Fig. 45. CORNICE SECTIONS

Theatre in Veszprem, Hungary. (See Fig. 196-197.) Note the lightness and curved outline characteristic of Ferro-Concrete.

Kaffeehandels A.-G., Bremen, Germany. This detail is typical of the "Wood-centring Style."

wood-mold superstition (Fig. 40-42, 45-55, 76B). Just as the intricate, delicate forms of late Gothic developed from the crude massiveness of the Romanesque, so the lace-like tracery and slender curved ribs of the Ferro-Concrete Style will replace the present primitive concrete shapes.

Many patents taken out relating to the use of concrete concern methods of casting, and the following figures give an idea of the range of possibilities in the forming of concrete. From 1890 till 1900 a total of thirty-five patents had been granted; in the years 1900-1908 one hundred and twenty-five patents; and between 1908 and 1914 eight hundred, or an average of three weekly. During the entire period (1890-1914) altogether nine hundred and sixty patents relative to concrete had been granted!

Structures bearing the stamp of the Ferro-Concrete Style—sculpture, as well as curved and slender forms that portray the plastic and tenuous qualities of reinforced concrete—have been created even with wooden forms. One of these is Unity Temple (Fig. 16, 17) on which its architect, Frank Lloyd Wright, makes the following comment:

"The slab-building is an expression of another method: Cast-slabs planned on multiples of 7'-0", set sidewise, and lengthwise, and flatwise make everything, as may be seen in the result. In Unity Temple the only limit was the mass of concrete that could withstand the violent changes of climate and remain related to human scale and easy construction. The box and blocks, however, determine the shape of every feature and every detail of the features, as it was all cast in "boxes". So a unit suitable for timber construction was adopted as the falsework in which it was cast was made of lumber. Multiples of 16", syncopated, was the scale adopted."[11]

The ribs of the Gothic church at Bois du Lac, Belgium, are strongly curved and the surface of the vault does not contain a single straight line

---
(11) "The Architectural Record", Jan. 1928.

Fig. 47. SCHOOL OF ARTS AND CRAFTS, HAMBURG, GERMANY
Staircase.
Prof. F. Schuhmacher, Architect.

Fig. 46. GEWERBEHAUS, HAMBURG, GERMANY
Lobby.
Prof. F. Schuhmacher, Architect.

Courtesy *Concrete*
Fig. 48a. CITY HALL, PASADENA, CAL.
Laminated curved forms. Due to location on the outside of a circular room, the arch has a concave face on the inside and a convex face on the outside.

Courtesy *Concrete*
Fig. 48b. CITY HALL, PASADENA, CAL.
Belt of laminated concrete formed by carefully tamping concrete with 7-in. slump in special curved laminated wooden molds.
BAKEWELL & BROWN, ARCHITECTS.

altho the centring was made of wood: boards 0.3 inches thick which were easily bent, supported by a series of wooden arches, formed the lower shell; the steep part of the vault was made without an upper centring-shell by carefully placing rapid-setting cement concrete at the bottom and then proceeding to the top.[12]

The complex vaulting in the hall of the Neue Sparkasse, Freiburg (Fig. 349) reveals the plasticity of "liquid stone" before setting. These vaults show what can be done with concrete by builders who set aside the popular prejudice that the new material demands crude, angular forms. When designing in the historical styles architects approve of curved outlines and delicate dimensions, but as soon as the new material liberates them from tradition, some architects show by continual use of shapes which require a minimum of thought and labor to design and build that their love of refinement is mere hypocrisy and then blame this vulgarity on concrete. Actually richer detailing of any material demands more thought on the part of the designer and greater expense. The plasticity of concrete facilitates the creation of curved and elaborate forms. There is no cutting away as with stone. When the mixture has a sufficient flow it will fill the most complicated mold including undercut parts (Fig. 48-55).

Wood-centring costs from thirty to fifty per cent of the total expenditure for concrete. The labor on the Bois du Lac vault amounted to fifty-eight per cent of the entire costs. The first step towards economy is to apply

(12) "Handbuch f. Eisenbet." Vol. IX, p. 213.

34

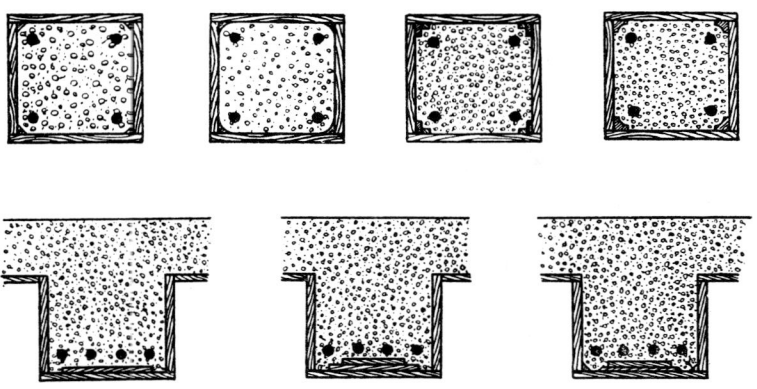

WOODEN MOLDS FOR FORMING PANELS AND PROFILES ON CONCRETE COLUMNS AND BEAMS

devices permitting the boards to be used again; eight systems of this type are listed in the *Handbuch für Eisenbetonbau.*[13]

Metal Forms.

Metal molds are frequently used, for they are easily handled and speed up production, and can also often be rented (Fig. 56). Some steel centring is adjustable and consists of elements that can be used for various curvatures. Braces have been used to adjust metal forms to different sizes. Rust-resisting steel sheets, which can be telescoped, are encircled by steel hoops, and employed to cast columns of varying heights and diameters. Metal forms are likewise utilized in various floor-systems. The fluting produced by corrugated steel pans has been left visible in some instances, which indicates the possibility of creating many kinds of textured concrete ceilings in a simple manner.[14]

The unity of a reinforced concrete structure, one of its most monumental qualities, is stressed by the smooth surface which metal forms produce. The form-marks preferred by some architects (Fig. 96) are characteristic of wood and a symbol of the "wood-centring style", but not a truthful expression of ferro-concrete.

Composition Molds.

Sand or sawdust, the former stiffened with cement, the latter with magnesium oxychlorid, can be formed into molds. The open balustrade, Fig. 57, was created by patty-pan cores of stiffened sand made from one standard mold which were placed on a table in the position of the voids in the balustrade. Each circle and connecting link was reinforced; after setting, the sandcores were knocked out.

Auto-Centring.

All the above types of molds have to be removed when the casting is finished. Permanent molds, self-centring reinforcement and self-supporting reinforcing utilize component

---

(13) *Hodges* centering consists of thin metal sheets nailed to lathstrips which in turn are fastened on two light metal chains. When rolled these mats can be carried by one or two men. By placing the mats alongside of each other and having them overlap vertically, stiffened by studs, walls of various dimensions can be poured.

(14) Climbing metal shuttering for walls is described by A. Lakeman in "Concrete Cottages, Bungalows and Garages", pp. 203-205. Edison's metal forms for casting a concrete dwelling in one day are described in the same book on p. 43, 44.

Courtesy *Concrete*
Fig. 49. CONCRETE BIRD BATH

Fig. 50. CONCRETE "TREES" AT THE 1925 EXPOSITION OF DECORATIVE ARTS, PARIS

parts of the finished structure as centring, whereby transportation costs, labor, material, and time are saved and the setting up of the molds is simplified.

PERMANENT MOLDS. "Building twice", the drawback of temporary forms discussed in connection with wood-centring, is eliminated by permanent molds. As everything that is set up is to remain, careful, accurate work is natural. Permanent centring consists occasionally of concrete slabs 0.8 inches thick. Columns are cast in precast hollow concrete cylinders which are sometimes reinforced. Architect Wielemans employed hollow concrete drums which bore sculpture on the outside. The lowest drum was set up and concrete filled in and tamped, then the second drum placed on top of that and so on till the column was complete. A. and G. Perret's famous concrete churches (Fig. 259-268) have ribbed columns which were probably cast in the same manner. Mr. Wielemans also cast domes on precast, ornamented concrete units acting as forms (Fig. 58). The advantages of this method are conspicuous. In another instance hollow drums of cinder-concrete, 1½ in. thick, were cast in metal forms. The inner centring consisted of a flexible core over which the spiral reinforcing and expanded metal lath were wound. After four hours the hollow drums were sufficiently set, and were then placed to form a hollow column into which concrete was poured. Mr. Frank Lloyd Wright's precast sculptured concrete slabs are a further example of permanent molds (Fig. 13-15, 163-174, 239, 240).

The two parallel ribs of the Belvidere bridge across the Kishuankee river, Ill., consist of concrete poured into hollow, precast concrete units placed so as to form the exposed sur-

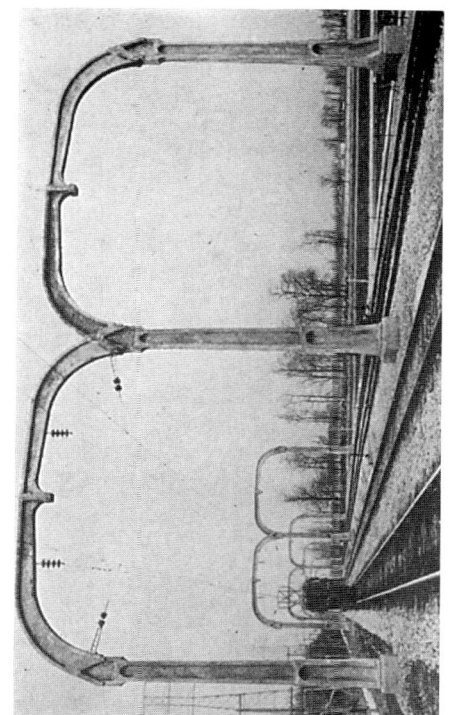

Fig. 52. FERRO-CONCRETE ARCHES OF HENRY FORD'S DETROIT, TOLEDO & IRONTON R. R.

Fig. 53. PAVILION FOR THE ST. HUBERTUS FOUNTAIN, MUNICH, GERMANY
The domes are of Ferro-Concrete.
AD. v. HILDEBRANDT, ARCHITECT.

Fig. 51. LAMP-MAST OF FERRO-CONCRETE IN THE AUSSTELLUNGSPARK, MUNICH, GERMANY
PROF. E. PFEIFFER, ARCHITECT.

Fig. 55. CLOISTER IN THE FRANCISCAN MONASTERY, WASHINGTON, D. C.

John J. Earley, Architectural Sculptor.

Fig. 54. MONUMENT IN VALHALLA MEMORIAL PARK, BURBANK, CAL.
Concrete-stone ornaments. Spanish tile on the dome.
W. T. Johnson and R. W. Snyder, Architects.

Courtesy *Concrete*
Fig. 56. EAGLES LODGE MILWAUKEE
Ribbed one-way metal mold floor, stained a mahogany color; the Craftex paint on the vaults was stippled

faces. An Austrian floor-system utilizes 1½ in. thick arched concrete shells which are laid on a thin concrete ceiling reinforced by wire, spaced to form narrow ribs; into these the concrete beams are poured. The Setz Floor, another European system, uses concrete planks as centring.

The cupola of the Zoological building in Cincinnati, O., was poured on a centring which combined features of self-centring reinforcement with permanent centring. Expanded metal was placed on a framework of steel angle arches and horizontal circular twisted bars. This network was plastered from below with a stucco of sand, gypsum, and cement, which hardened and served with the reinforcing as a mold for the 3.9-in. thick concrete shell which was then poured. Professor Mecenseffy's book mentions three domes on German buildings, poured without centring, which probably were cast in a similar manner. Professor Dom. Bohm made all the vaults of the Neu-Ulm church (Fig. 320-331), except those of the Baptis-

Courtesy *Concrete*
Fig. 57. CONCRETE BALUSTRADE, YONKERS, N. Y.
H. V. WALSH, ARCHITECT.

tery and the Resurrection chapel, with permanent molds by the Bohm-System: The main reinforcing rods are fastened to temporary wooden joists by wires. The remaining reinforcing rods are in turn attached to the main bars. A mesh of small burnt-clay elements united by wire—*Stauss-bricklets*—is tied to the reinforcing network and a concrete layer of one inch, which must not be too wet, applied on top of the mesh. Concrete of one part Portland Cement, one part clean river sand and one or two parts pumice-sand or volcano-ash is best, due to its light weight and acoustic properties. Two or three days later ordinary concrete is tamped or poured from above in the required thickness and the *Stauss* mesh plastered with cement mortar from

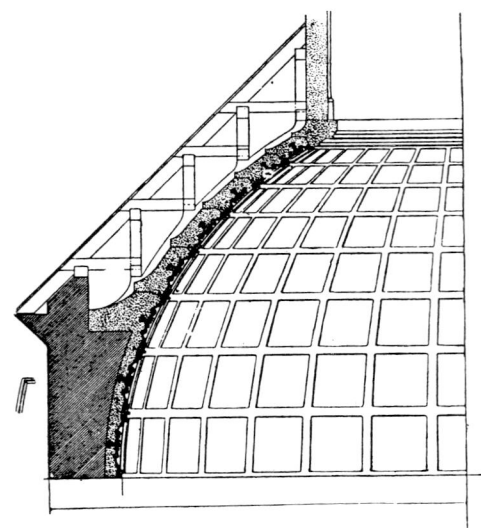 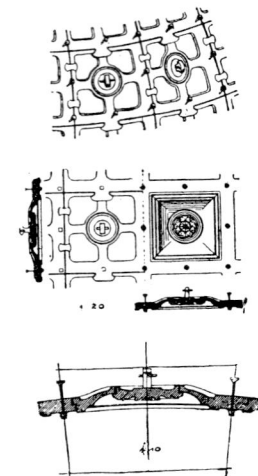

Fig. 58. PRECAST CONCRETE ORNAMENTS ACTING AS PERMANENT MOLDS FOR A CONCRETE CUPOLA

The concrete was placed around the bottom ring of coffers first and then further layers poured.

VON WIELEMANS, ARCHITECT.

Fig. 59. REINFORCED CONCRETE GRILLE IN TEMPLE EMANU-EL, SAN FRANCISCO, CAL.
J. BAKEWELL, JR. & A. BROWN, JR., ARCHITECTS.

Fig. 60. TEMPLE EMANU-EL, SAN FRANCISCO, CAL.
Concrete Grilles reinforced with heavy galvanized wires.

below. After the concrete has set the supporting wires are cut and the wooden joists removed.

The casting of the dome-windows in the Antwerp Station, Belgium, is an interesting example of the use of permanent molds. The arcaded part was poured in ordinary forms placed in the wall. The negative of one side of the large semi-circular part was laid out on a horizontal platform. The mold of this ornamented tracery consisted of a plastic mass covered by a paste of magnesium oxychlorid and sawdust in a thickness of from one to two inches; it set quickly and gave a light and firm mold which could be treated like wood. Concrete was poured into it, forming a two-inch layer which was then cut into sections. The magnesite mold was used repeatedly to create a series of units which, when placed back to back, formed a permanent centring, the exposed surface. Into this permanent mold of concrete, reinforcing was placed, and concrete poured.[15]

The grilles of the Temple Emanu-El in San Francisco (Fig. 58-60) were cast in sections, using a dry mix, the sections being made with hollow backs; the reinforcing was looped out at places into the grooves in the back. When these sections were assembled a system of wiring was placed in the grooves forming an exposed steel grille at the back tied together and to the loops of the reinforcing of the sections. Then the whole back was filled with grout forming a light steel grille bedded in concrete.

(15) **Twelvetrees**, "Concrete-Steel Buildings", 1907.

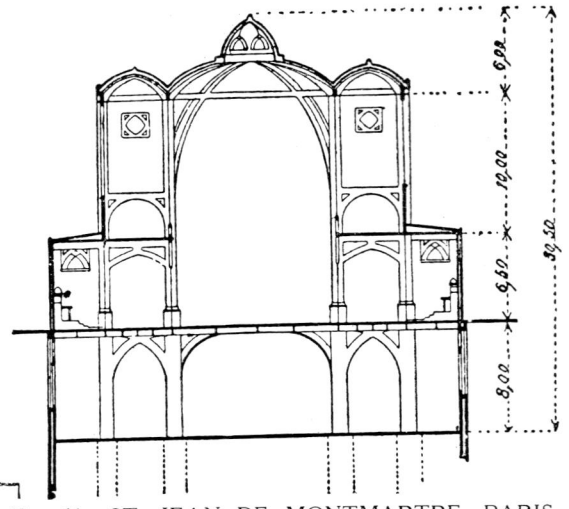
Fig. 61. ST. JEAN DE MONTMARTRE, PARIS
Transverse Section.
DE BAUDOT, ARCHITECT.

Various kinds of veneer have been used as permanent centring; they should have a rough inner side so as to unite more firmly with the concrete backing. Slabs of stone, marble or terra cotta which are anchored in the concrete serve as molds. These slabs must be safeguarded against the pressure of the fresh liquid concrete. One of the earliest examples of brickmolds is Baudot's church, St. Jean de Montmartre, Paris (Fig. 61). The horizontal and curved slabs and the cores of the column-tiles are of concrete; all other parts consist of hollow tile with reinforcing bars. The *Poyet Floor* utilizes special hollow blocks and quick-setting cement to create a ceiling which serves as centring. In other instances units of gypsum, hollow glass-stones[16] and glass panes have been used as permanent mold. An original type of centring is used by the *Zibell System* for concrete floors: cardboard "pipes" stiffened by removable steel spirals are inserted into the main forms and create voids which reduce the weight.

---
(16) See **Decorative Inserts** Chap. II.

SELF-SUPPORTING REINFORCEMENT. The reinforcement of these systems is built up as a rigid unit which carries the centring and the fresh concrete. Structural steel is often used for this purpose. Some European floor systems utilize the reinforcing steel as a beam that temporarily carries the centring. The central dome of the New Anatomy Institute, Munich, had rolled steel and reinforcing bars forming a skeleton, which was covered with metal lath and carried the concrete shell before the cement set. The reinforcement was dimensioned sufficiently to carry the weight of the cupola, the centring, and the workmen (Fig. 238). A similar method was used in building the Christian Science Church in Los Angeles.

SELF-CENTRING REINFORCEMENT. Expanded metal of such gauge that the concrete cannot "run" thru, and wide enough to let it ooze thru and envelop the metal is called self-centring (Fig. 62-64). Wire mesh carrying waterproof paper backing fulfills the same purpose (Fig. 65). Self-centring reinforcement seems to be the highest type of centring since it combines to some extent the advantages of permanent centring with those of self-supporting reinforcing. The proper consistency of the concrete is essential for the success of this method. Self-centring is made possible by the fact that the reinforcing has to be placed in the tension zone which is near the surface usually. Several firms market self-centring expanded metal, which not only eliminates wood centring but far surpasses it by being bendable. Self-centring is pliable in one direction. U or T-shaped ribs

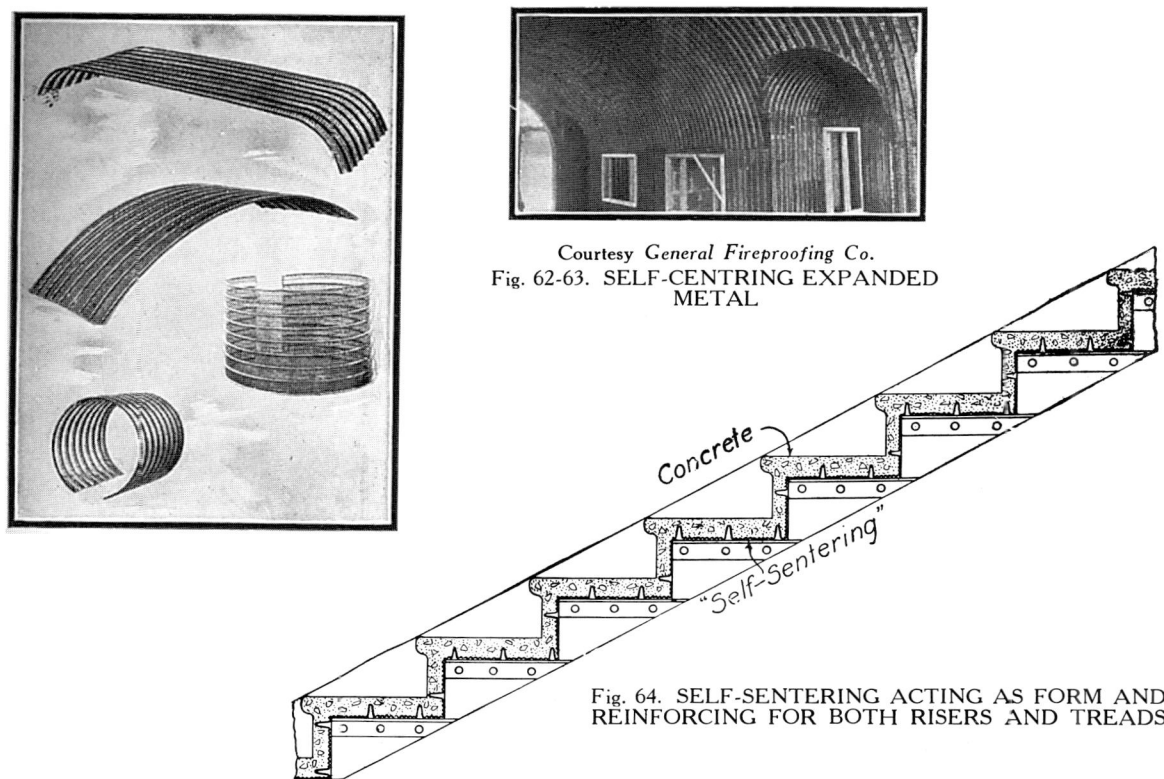

Courtesy *General Fireproofing Co.*
Fig. 62-63. SELF-CENTRING EXPANDED METAL

Fig. 64. SELF-SENTERING ACTING AS FORM AND REINFORCING FOR BOTH RISERS AND TREADS

pressed in the expanded metal give rigidity and a supporting network of bars becomes unnecessary: the erection of the centring and the placing of the reinforcement is *one* process. The sectional area of the reinforcing can be increased by adding light bars. The stiffening ribs make expanded metal very rigid lengthwise of the sheet, but by means of a bending machine self-centring can be curved to any desired radius, twelve inches or more; it can be cut in every direction. By bending the edges, delicately curved boundary lines can be created; wall apertures with curved outline can be cut out of it. The "wood-centring style" gives way to the Ferro-Concrete Style.

In erecting a factory in Brooklyn hollow cylinders of self-centring expanded metal to which reinforcing bars had been attached by wires, were fastened to rods projecting from the foundation walls and the concrete poured from above. One manufacturer delivers the lath in sheets, which, interlocking, form the two mold-shells and act also as reinforcing. A quaking mixture of $1:2\frac{1}{2}:3\frac{1}{2}$ when tamped spreads thru the mesh overlapping the lath on the outside in a layer about $\frac{1}{2}$ in. thick. A 1:2 cement mortar is frequently plastered on the interior and exterior before the concrete sets so that it becomes part of the monolith.

Methods of Placement.

Concrete can be placed in various ways. For buildings covering large areas, sometimes tracks are laid on which the concreting units run. The Breslau Centenary Hall (Fig. 380-384), for example, was built by lay-

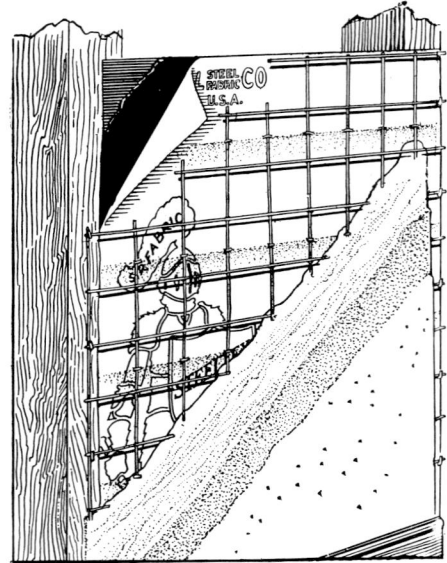

Courtesy *National Steel Fabric Co.*
Fig. 65. STEELTEX — WIRE MESH WITH WATERPROOF PAPER BACKING WHICH ACTS AS REINFORCEMENT AND CENTRING

ing a circular track of two hundred meters diameter on which travellers ran around a large central tower; overhead cranes connected these towers so that every part of the building site could be reached.

In the gravity system concrete is hauled up a central tower and drops thru chutes to any part of the building desired.

Compressed air is used by different systems. In some it forces the already mixed mass thru a hose to the centring or throws it on the surface which is to be covered. The cement-gun and similar machines throw the dry sand and cement on the centring while a jet of water sprays the mix on its flight (Fig. 66). Gunite, the resulting concrete, is denser than that laid by hand. Tests show that it is from 20 to 70 per cent stronger than concrete placed by hand, and less permeable. The cement-gun reduces the molds to one shell, as only a backing needs to be placed behind the reinforcing mesh. Layer after layer can be shot so that walls of any thickness can be produced. Dwellings have been built by this method with a backing of cheesecloth supported by studs as centring. Walls consisting of two or three shells with insulating airspaces are easily made, as the cement-gun is best adapted to build thin shells of concrete.

Courtesy *Cement Gun Co.*
Fig. 66. "SHOOTING" CONCRETE WITH THE CEMENT GUN

Fig. 67-69. ERECTION OF HOUSES BY THE ATTERBURY STANDARDIZED SECTIONAL SYSTEM, FOREST HILLS GARDENS, L. I.
GROSVENOR ATTERBURY, ARCHITECT.

Courtesy *Concrete*
Fig. 70. GRAUMAN'S CHINESE THEATRE, HOLLYWOOD, CAL.
Reinforced Concrete Framework.
MEYER & HOLLER, ARCHITECTS.

Shooting concrete on self-centring, self-supporting reinforcement backed by a light material would seem to be the ideal method of creating concrete structures. Centering costs are eliminated and every conceivable shape and form can be easily produced, while complete freedom to fulfill all possibilities of ferro-concrete is achieved. Two domes erected for the Zeissworks, Jena, Germany, with diameters of 52'-6" and 131', received self-supporting reinforcing consisting of flat bars two feet long,

---

(17) "Concrete and Constructional Engineering", May 1927, p. 309. In "Engineering News", Vol. 74, p. 4 the erection of a wall by shooting gunite on self-centring reinforcement is described.

Courtesy *Concrete*
Fig. 71. PERMANENT RECORDS IMPRINTED IN SOFT CONCRETE. GRAUMAN'S CHINESE THEATRE

curved to conform to the shape of the dome, which were connected by two steel discs and bolted together. The assembling of the reinforcing was begun from the top and when completed was covered with wire mesh against which concrete was placed by a cement-gun operated by men working in cradles suspended from the reinforcement.[17]

Fig. 72. GRAUMAN'S CHINESE THEATRE, HOLLYWOOD

Courtesy *Concrete*
Fig. 73. GRAUMAN'S CHINESE THEATRE, HOLLYWOOD
Concrete-stone ornament and colored cement stucco.

Temporary centring satisfies the needs of the engineer in producing structural forms; the cement-gun and self-centring reinforcement enable the architect to produce those delicate and intricate elements, curved surfaces, and ornaments which represent his language in durable concrete. Experimentation will probably show that many decorative possibilities lie latent in the "cementgun-selfcentring" method.[18]

Centrifugal force has been used to make hollow concrete blocks. Hollow masts have been produced by letting concrete rotate in a hollow cylinder, thus throwing it by centrifugal force to the shell where it covers the reinforcement. Concrete can be molded on wire mesh frames (Fig. 1) and formed with templates; this method can also be combined with the use of cast parts. R. C. Davison gives detailed instructions covering these methods in his book "Concrete Pottery and Garden Furniture".[19]

Cylindrical columns of concrete have been shaped by turning and the R. M. Jones system achieves similar results by pouring concrete on an

---
(18) See last page of Chap. II.
(19) Munn & Co., New York, 1910.

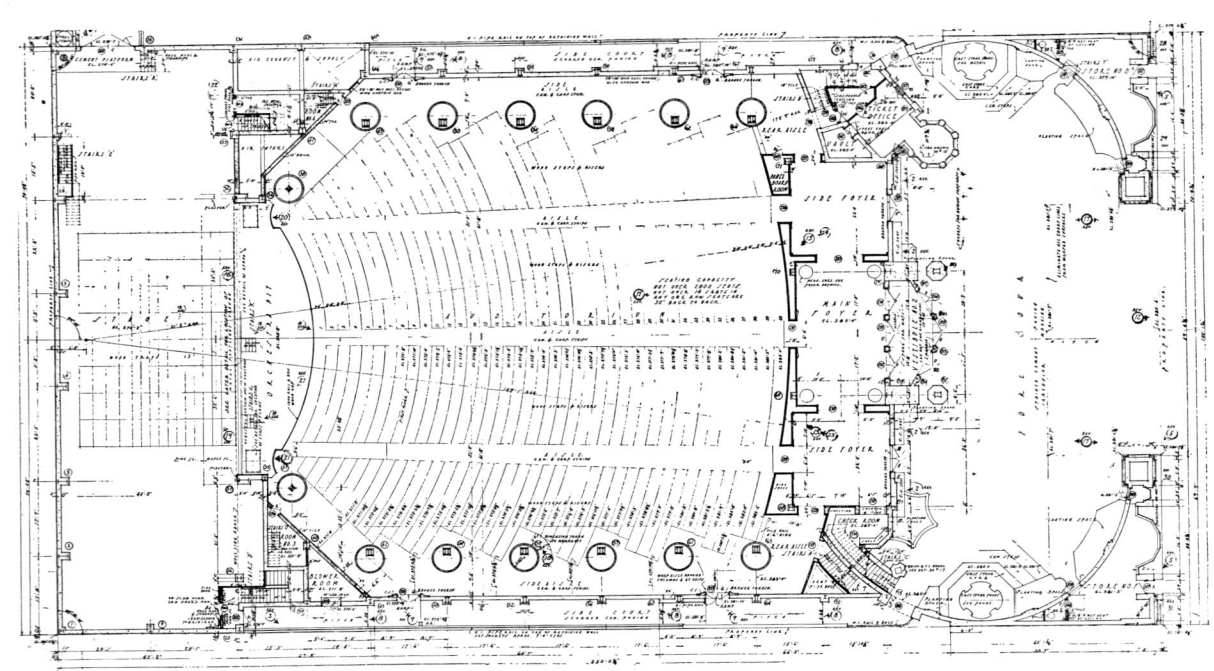

Fig. 74. GRAUMAN'S CHINESE THEATRE
PLAN of Auditorium.

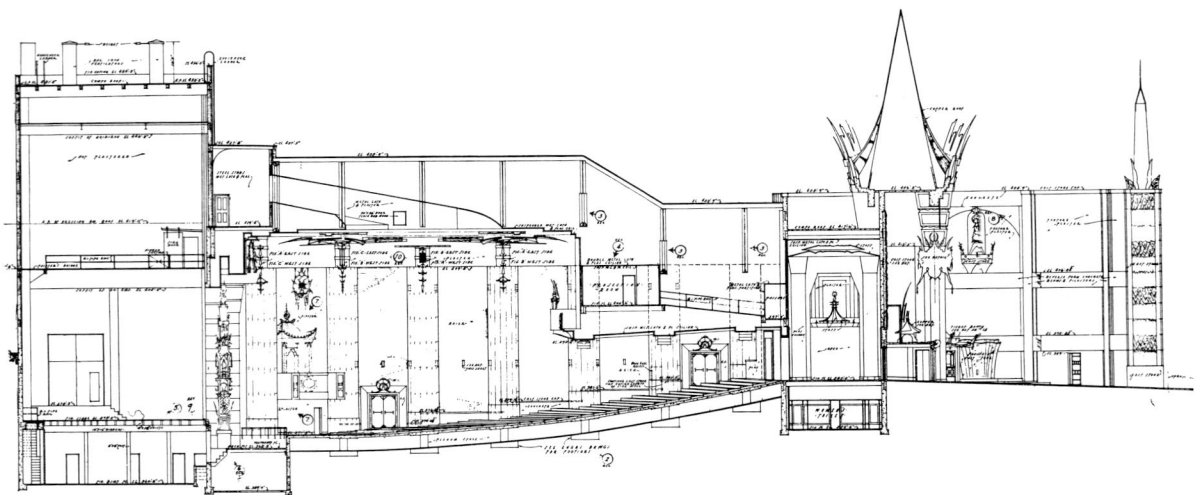

Fig. 75. GRAUMAN'S CHINESE THEATRE
Longitudinal SECTION.
MEYER & HOLLER, ARCHITECTS.

Fig. 76a. CONCRETE CEILING CAST IN GLUEMOLDS.
SAVINGS BANK, SACHSISCHES HOUSE, DRESDEN, GERMANY

endless band which runs thru a system of pins.

The production of concrete structures is variable according to the place of casting. Usually the structure is cast on the site. Sometimes all elements are cast in the shop and, when set, transported to the building-site. Mr. Grosvenor Atterbury employed this system for workingmen's dwellings (Fig. 67-69); the units filling the spaces between windows and doors weighed 7 or 8 tons and small rooms were covered with a single slab. A third method consists in a combination of the first two systems. The bulk of the required aggregates is often available near the building site which causes a reduction of transportation costs.

As the described methods show, wood-centring is only one of many methods for forming concrete and one which really limits the possibilities of "liquid stone". Reinforced concrete combines the advantages of steel, stone, and wood, eliminating many of the disadvantages of each. The reinforcing bars, bendable and not limited as to length and slenderness, can follow the most complex lines and curves and give the surrounding concrete an indestructible backbone. Being plastic when placed, concrete can adopt any conceivable form. Practically unlimited length, slenderness, and intricacy are combined with strength and durability in concrete. Fig. 70-75, 77 show a striking example of these qualities. Ferro-concrete can project in any direction from the supporting member, following a straight or a curved line. In contrast wood is of limited length and width and requires complicated connections; curved forms necessitate much labor. Stone and brick offer actually considerable re-

Fig. 76b. FERRO-CONCRETE GRILLE
Von Wielemans, Architect.

sistance thru their limitation in size, the difficulty of cutting, and the need of anchoring. Concrete is stronger than terra cotta and still has the possibilities of clay due to its plasticity. That the historical styles show such a marvelous wealth of intricate ornament is due to the passionate love of beauty which lived in the old masters; they created their tracery and finials in spite of the brittle and limited materials at their command. Steel itself is too skeletonlike to be architectural; its virtues are incorporated in ferro-concrete which eliminates gusset-plates, rivets, and bolts and creates a structure which is a unit from top to bottom. Soft, curved ceramic forms, Moorish stalactites, as well as the passionate architecture of the Baroque, can be made in durable concrete without conflicting with its characteristics. Concrete can be cast in metal, gypsum, or glue molds and the finished product will be exquisite sculpture (Fig. 76, 128-179).

Ferro-concrete, treated like Cinderella by most contemporary architects, in reality combines the possibilities of casting and of sculpturing and thru its reinforcing steel can partake of some characteristics of the wrought-iron technique. D. C. Allison states the problem clearly in saying:

"We have not realized that Cinderella responds just as readily to a little attention, and a little loving as do her sisters such as stone, marble and terra cotta. We have as a mater of course, spent unlimited energy in working, carving, beautifying these other materials....Concrete however, we have hesitated to handle more gently or more intimately than could be done by means of a wheelbarrow, a shovel and a mixer. The one thing most needed is for more architects of designing ability

Fig. 77. TOWER CARRIED BY THREE CONVERGING HALF-ARCHES OF FERRO-CONCRETE. FAIRGROUNDS, COLOGNE, GERMANY
Wahl & Roder, Architects

to realize that this material is capable of unlimited development; it can be molded into any form that the imagination can conceive, knit into the very fibre of a structure—an integral homogeneous part of it, and may frankly be brought clear through to the surface and admit its identity, honestly, convincingly, beautifully. If we are willing to spend but a fraction of the cost of carving and working granite, stone and marble, upon the building of plaster molds or in ornamenting surfaces with scraffito, or stucco in its many forms, absolutely any degree of architectural richness desired may be obtained, and that at a cost very much less than in any other material of like permanency."[20]

When once the wood-centring superstition has been banished from the architect's studio, the unlimited forming possibilities of ferro-concrete will incite the designer to greater daring.[21]

---

(20) "Proceedings of the Am. Conc. Inst.", 1926.

(21) The fifth group of ferro-concrete's possibilities—those of surface treatment and sculpture—is discussed in the following chapter.

Courtesy *Concrete*
CONCRETE BRIDGE, BECKENRIDGE PARK, SAN ANTONIO, TEXAS
Formed without molds by modelling.
DIONICIO RODRIGUEZ, SCULPTOR.

CHAPTER II
SURFACE TREATMENT AND SCULPTURE

Fig. 78. SHAFT HEAD HOUSE, ISHPEMING, MICH.

"Climatic conditions in this country, particularly in the spring and fall, when frost alternates with warm temperatures, are severe on all kinds of concrete construction, but provided the concrete is well made it should suffer no greater disintegration than the hardest stone exposed in a similar manner."—*Oswald C. Hering*.

Courtesy *Concrete*

Fig. 79. ST. JOHN'S EPISCOPAL CHURCH, LOS ANGELES, CAL.
Monolithic Concrete.

Although concrete permits the greatest variety of surface treatments, many architects continue to hide it behind brick or stone facing. The appearance of the surface depends on the care exerted: The bad reputation which concrete surfaces have is really due to the lack of ability of the builders and workmen and not to concrete itself. John J. Earley writes of this paradoxical situation:

".... All of these things when not controlled by a proper technique naturally cause the form of concrete to fall short of architectural requirements. These shortcomings have unfortunately been so general, indeed, that architects have come to accept them as necessary and have allowed modifications of form, have omitted architectural moldings and reliefs because fitting execution could not be expected in concrete. Contractors have encouraged them in this belief and have by this means avoided technical difficulties. Some of the experiences of my studio would be

Fig. 80. EFFECT OF WIRE BRUSHES AND WATER ON A CONCRETE SURFACE-LAYER *(1 p. Cement, 2 p. River-Sand, 3 p. White Gravel).*

humorous if they were not sad. Architects submit their problems of concrete surface treatment with apologies, with fear that their drawings are too elaborate even though simplified to the point of crudeness, and with hope that we will do our best to retain what little is left of the esthetic. The color and texture of concrete are in a similar state of disorder and from a similar cause, namely misapplied technique."

Hundreds of concrete buildings all over the world—in cold Finland as well as in tropical Florida—testify that proper workmanship produces durable, pleasing concrete surfaces. Brick, stone, and marble backed by traditions of thousands of years are not so exposed to slovenly treatment as is their modern cinderella-sister, concrete.

Fig. 81-82. TOOLED CONCRETE SURFACES (GERMAN MUSEUM, MUNICH, GERMANY)

PRIMARY TREATMENTS
Form-Marks.

Chipping off the "fins" left at the board-joints and filling the voids is the simplest treatment. Oiled paper or metal sheathing placed in the molds insure a smooth surface. Further the boards can be planed, or covered with a coating of lime or spirit-varnish which prevents the concrete from sticking to the mold. The board-joints can be filled with clay or mortar or covered with strips of paper or cloth. Mr. Twelvetrees recommends a liberal coating of oil against which fine sand is blown. He further suggests the removing of efflorescence by washing with plain water and of hair cracks by brushing the cracked sur-

Fig. 83. COKE-BIN OF THE WENDEL MINE, GERMANY

Fig. 84. CONCRETE KIOSK, FOREST HILLS GARDENS, L. I.
The roof is precast blue aggregate.
GROSVENOR ATTERBURY, ARCHITECT.
S. PHELPS AND J. TOMPKINS, ASSOCIATED.

Courtesy *Concrete*
Fig. 85. CLUBHOUSE OF THE COUNTRY CLUB, WHEELING, ILL.
Hydraulic Art Stone faced with a mixture of pink and black granite and a small amount of carborundum with white waterproof cement.

Courtesy *Concrete*
Fig. 86. LOBBY OF THE COUNTRY CLUB, WHEELING, ILL.
Concrete blocks with a facing of 60% white marble, 20% silica sand and 20% dark banksand with white waterproof cement; they resemble limestone. The baseboard has a black concrete facing.

face; both defacements can be eliminated by washing with a weak solution of hydrochloric acid. To remove the grain marks of wood forms a creamy mixture of neat cement and water is applied with a brush and preferably rubbed in with carborundum stone.

The churches of A. and G. Perret (Fig. 259-268) show the form-marks, but probably for the sake of economy. A concrete structure is a monolithic cube with voids cut out (Fig. 224); form-marks contradict this glory of ferro-concrete by reminding of the limited pieces that make the molds for unlimited concrete. Electric driven concrete surfacing machines are sometimes used to remove the marks left by the centering.

Exposing the Aggregate.

The monotonous grey cement-film which covers the concrete surface can be removed by scrubbing with a bristle or wire brush and clear water within twenty-four hours after pouring the concrete (Fig. 80). To be able to do this while the concrete is yet green the forms must be removed as soon as the concrete is stiff enough to carry its weight and so only two or three feet are placed at a time. The concrete must be allowed to attain sufficient set so that the particles of aggregate will not be disturbed. Sand-

JOHN J. EARLEY, *Architectural Sculptor* MURPHY & OLMSTED, *Architects*

SHRINE OF THE SACRED HEART, WASHINGTON, D. C.
IN THIS APSE, ALL OF THE ORNAMENTATION, INCLUDING FIGURES, IS CARRIED OUT IN CONCRETE MADE WITH ATLAS PORTLAND CEMENT AND EXPOSED COLORED AGGREGATES.
"Never once did concrete fail us"— John J. Earley

Fig. 87. SHRINE OF THE SACRED HEART, WASHINGTON, D. C.
Exposed aggregate concrete prevents the cement-water paste from dominating the finished appearance.
J. J. EARLEY, ARCHITECTURAL SCULPTOR.
MURPHY AND OLMSTED, ARCHITECTS.

Fig. 88. CAPITAL, SHRINE OF THE SACRED HEART

J. J. Earley's technique causes the aggregates to be the dominant surface elements and controls the characters and positions of the various colored particles. (Compare the illustration on the jacket and the frontispiece).

Fig. 89. CONCRETE ORNAMENT IN THE SHRINE OF THE SACRED HEART
Vari-colored aggregate.

blasts are practical for treating large buildings and are valuable for hardened concrete; they can remove the cement-film from quite thin facing layers which might be damaged by a hand-pick.

R. F. Havlik recommends that the surface of the green pre-cast product be treated with a very fine spray of water under pressure just as soon as it is removed from the mold. This washes off the surface-cement and drives it into the concrete, thus exposing the natural surface of the aggregate. The product is then cured until it is hard. Sometimes it is further treated by scrubbing it with a solution of 1 part muriatic acid to 2-5 parts water. No dust or fine material below about a Nr. 20 mesh should be mixed into the surface-layer if checking is to remain invisible.[22] Chemicals which remove the cement-film are diluted hydrochloric (muriatic) acid and sulphuric acid, but when the aggregate contains $CaCO_3$ they should not be applied. All these treatments must be made before the concrete has fully set and to avoid discoloration they must be washed off thoroughly. The outer film can also be removed by rubbing down the surface with a piece of stone, and where the aggregate is soft stone, a terazzo finish is achieved. In one case wooden floats were used to remove the joint-marks, the final rubbing being applied with carborundum and water. Cass Gilbert writes of a concrete bridge: "The aggregate was of a trap rock or granite that gave a rich, warm color and a beautiful texture to the surface that could scarcely be rivaled by any other material."[23]

Textures as shown in Fig. 81, 82 can be obtained by tooling concerning which W. W. Clifford writes:

"Tools cannot be used until the concrete is thoroughly hardened. When a hand pick is used a comparatively large amount of concrete is scaled off leaving a coarse textured finish. If the pick is used at right angles no lines or marks are left. By striking a glancing blow, tooth marks are left which may be made parallel or at various angles. Bush hammers give a surface similar to that produced by the pick but with a finer texture. Four to eight-cut hammers are used with similar results to those obtained on natural stone."[24]

In one instance concrete blocks of which the outer surface to a depth of one inch was composed of white cement, marble or granite aggregates, and an integral waterproofing compound

---

(22) "Concrete", March 1923.
(23) "The Architectural Forum", Sept. 1923, p. 84.
(24) "The Architectural Forum", Feb. 1922, p. 69.

ITALIAN
Random troweling reduces
the irregularities of this rough-torn
surface

ITALIAN BRUSH
A smoother, more carefully finished
example of Italian textures

ENGLISH COTTAGE
A plastic surface coat made irregular
with random strokes of a broad,
soft-bristle brush

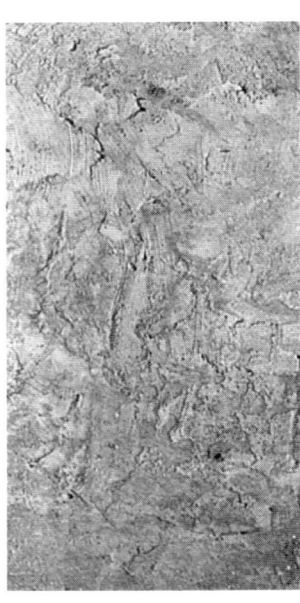

FRENCH
Controlled manipulation of a wood
float produces a surface of pleasing
irregularity

FRENCH BRUSH
A brush is used to soften the
sweeping, semi-circular
trowel marks

GOTHIC
A floated finish rough-torn with the
back edge of the trowel

Fig. 90. STUCCO TEXTURES

Fig. 91. AL MALAIKAH TEMPLE, LOS ANGELES, CAL.
Reinforced concrete with stucco finish.
J. C. Austin, Architect.

were treated with a compressed air tooling machine which brought out the glisten of the aggregates.

COLORED CONCRETE

Cements from different parts of the country, although of the same brand, differ in shade. Colored aggregate and those mineral pigments which may be safely added to the cement in small quantities (up to ten per cent) are suitable for coloring exterior surfaces. J. J. Earley points out that even white cement has sufficient color to interfere with that of an added pigment, and that as only a small quantity of pigment can be added, it can not give concrete a hue of great strength as required for adequate decoration. Ralph C. Davison states:[25]

---

(25) "Concrete Pottery and Garden Furniture", pp. 94, 95.

Courtesy *American Architect*
Fig. 92. AL MALAIKAH TEMPLE—STAGE
The proscenium opening has a clear span of 100 ft. and a height of 37 ft. at the crown. Amplifying speakers are concealed in the jeweled crown of the arch.
The frescos in the spandrels are painted on canvas and then applied to a perfectly smooth plastered surface. The fresco work is done in lead and oil paint. The plastering on the spandrel is directly on the concrete, no furring being used.
J. C. AUSTIN, ARCHITECT.

"By mixing mineral pigments, known as lime or cement-proof colors, with white Portland cement it is possible to produce shades which will be absolutely permanent. They come in powdered form and should be mixed with the dry cement and marble dust or white sand until the whole mass has a uniform tint...

By weighing the amount of pigment used, a much more uniform result can be obtained than by judging the amount by bulk. By mixing red, yellow and blue every tone may be obtained. By making small test pieces of mortar, and noting the color after they are dry, the proper amount of pigment can be determined.

The following pigments which can be procured in dry colors give satisfactory results:

Fig. 94. WILSHIRE BOULEVARD CONGREGATIONAL CHURCH, LOS ANGELES
Precast concrete was used at the entrance doors and for the colonnettes, but most of the remaining decorations were cast monolithically by plaster molds and by nailing blocks in various patterns on the inside of the forms before concrete was poured.
ALLISON AND ALLISON, ARCHITECTS.

Fig. 95. UNIVERSITY CLUB, LOS ANGELES
Entrance.
The detail is executed in precast concrete-stone.
ALLISON AND ALLISON, ARCHITECTS.

Fig. 96. PUBLIC LIBRARY, LOS ANGELES
Cement-stucco over reinforced concrete.
B. G. Goodhue, Associates and C. M. Winslow, Architects.

Dry Pigments	Resulting Color
Red oxide of iron	} Red
Venetian red	
White Portland Cement	White
Ultramarine blue	} Blue
Oxide of cobalt	
Chromate of lead	} Yellow
Yellow ocher	
Chrome oxide of copper	Green, light
Carbonate of copper	Green, dark
Burnt umber	Brown
Ordinary Portland Cement	Gray
Lampblack	} Gray or black
Torch black	(according to
Black oxide of copper	quantity used)"

W. W. Clifford writes: "It can hardly be said that even any of the mineral colors are permanent under severe exposure"[26] and O. C. Hering advises more daring in selecting the shades and tones as "the action of sun, wind and water conspire to bleach, and the dust to blacken, the chaste tones so fondly conceived". He further writes: "Mixing in paste form is sure to distribute the color unevenly. By mixing the ingredients dry, a sufficient amount for the entire operation may be obtained and kept until needed, pure water being added to such quantities as may be immediately required which prevents unevenness in color."[27]

(26) "The Architectural Forum", Feb. 1922, p. 69.
(27) "Concrete and Stucco Houses", pp. 31, 26.

Fig. 97. MEMORIAL SHAFT, ASTORIA, OREGON

The scraffito was created by superimposed thin layers of concrete of different tints cut thru to the correct depth to complete the design in the desired colorings.

Colored Aggregate.

Facing layers consisting of finer aggregate have colors varying according to the kind of aggregate selected (Fig. 83-85). The tone of the surface can be materially changed for special parts of the work by adding a small amount of aggregate of contrasting color, such as white or yellow marble to give a lighter tint, or black marble, iron slag or blue granite to give a darker shade. The surface-layers are packed with the coarse material by using a sheet metal dam with angles riveted on to keep it at the proper distance from the face of the mold; it is removed before either mixture has set. A second method consists in spading the concrete back from the front of the forms and introducing the finishing mixtures, tamping it with the rest of the wall. The thickness of surface mixtures placed in this way will be 1 to 2 inches. Surface mixtures may also be plastered onto concrete or other suitable walls (stuccowork), or they may be poured 3 to 4 inches thick against the backing wall which has projecting wires for bonding. In precast concrete the facing mixture is placed first and the backing poured on top of it.

After crushing, the aggregates are screened and graded into not less than 3 sizes ranging usually from $\frac{1}{4}$ to $\frac{1}{2}$ inches. Then only those particles which are of the sizes that will give the desired texture are recombined in

Fig. 98. PUBLIC LIBRARY, LOS ANGELES
Detail of Rotunda Dome.
The designs are painted on concrete.
BERTRAM B. GOODHUE, ASSOCIATES AND C. M. WINSLOW, ARCHITECTS.

Fig. 99. PUBLIC LIBRARY, LOS ANGELES
Ceiling Detail.
J. E. GARNSEY AND A. W. PARSONS, MURAL PAINTERS.

Fig. 100. UNIVERSITY CLUB, LOS ANGELES
Ceiling Detail.
The concrete was cast in metal forms and then painted.
ALLISON AND ALLISON, ARCHITECTS.

proportions that make a remarkably dense and uniform mixture in which the voids constitute less than twenty per cent of the volume. To this recombined mass of aggregates, portland cement and water are added in those proportions which will give a workable mix and which will, at the same time give a strength at 28 days of approximately 3000 pounds per square inch. Some water is extracted from the concrete after it is deposited in the molds. The present practice is to leave in the concrete at this stage about four gallons of water for each sack of cement used in the mix. The removal of the excess water, which was used as a vehicle for placing, increases the density, strength and impermeability, and consequently the durability, of the concrete and diminishes the shrinkage or tendency to draw away from the molds during the process of hardening. The forms are removed before the concrete has hardened too much and the cement-film is removed by one of the methods already described.[28]

The various veneers used to cover brick and rubble can fall off, but the concrete facing layer is part of the monolithic concrete structure. Colored cement, clay, glass, metal chips, mosaic cubes, glimmer and mother of pearl can be mixed into the facing ag-

---

(28) A specification of exposed aggregate work is reprinted in "The Architectural Forum", Feb. 1922, p. 70.

69

Fig. 101. EDGEWATER BEACH HOTEL, CHICAGO
West Lounge Room.
Painted concrete beams.
MARSHALL AND FOX, ARCHITECTS.

Fig. 102. UNION TEMPLE, BROOKLYN, N. Y.
Main Auditorium seen from the front.
ARNOLD W. BRUNNER ASSOCIATES, ARCHITECTS.
NATHAN C. JOHNSON, CHEMICAL CONSULTANT FOR THE PAINTED CEILING.

gregates to create a colorful surface (Fig. 86). White gravel mixed with white portland cement produces a gleaming white surface. Another attractive surface layer can be created with crushed granite and glimmer. One German firm applies cement mortar with pulverized colored ceramics and glass-bits called "Keramic Kristall" stucco.

John J. J. Earley, the master-craftsman, gives the following description of his method[29] (Fig. 87, 89, the jacket and the two colored plates):

"The elements of a good appearance are three, namely, good form, color and texture. And the experience of the past few years has taught that concrete of any form, color, and texture can be made by the proper selection of the materials, particularly of the aggregates and by the effective control of them while in a plastic state. With regard only to its appearance let me suggest that concrete be thought of as an aggregate, which is held in place by the least possible amount of hardened cement paste, and which before the hardening of the cement was flowed into place in a vehicle of water. This gives the idea that it is the aggregate which takes the form and gives the color and texture, that the cement is a binary material and has no part in the appearance, and that the water is a carrier which places the material with the least amount of work. A peculiarity of concrete is that the cement-water paste, which in the relation of its volume to that of the aggregate constitutes but a minor

---

(29) "Substance, Form and Color Through Concrete" and other articles by Mr. Earley.

Fig. 103-104. UNION TEMPLE, BROOKLYN, N. Y.
Details of painted concrete ceiling in the Main Auditorium.
W. Gehron, S. F. Ross & M. W. Alley, Architects.

Fig. 105. NEW MESS HALL, WESTPOINT, N. Y.
Historical scenes are painted on the concrete beams.
A. W. Brunner Associates—W. Gehron, S. F. Ross & M. W. Alley, Architects.

part of the mass, dominates the appearance of the whole ... the aggregate is the dominant element of concrete, therefore it should exercise a dominating influence on the structure and appearance ... It is by constructing a skeleton of aggregate that volume-changes, segregations, and settlement are prevented. It is by causing the aggregate to occupy a very great part of the surface that predetermined color and texture are obtained."

Describing his work on the Shrine of the Sacred Heart in Washington, D. C., Mr. Earley continues:

"... In reality, concrete, because of its plastic nature and because of a technique which makes the aggregate take the form and produce the color and texture, was able to assume as many hues and values as there were aggregates to produce them. The same marbles as were used in the ancient churches could be used as aggregates in modern concrete; many stones of great decorative value, but too hard to be worked, were available for concrete; mosaic materials of all kinds could be used in concrete without serious loss in color ...... By considering the particles of aggregate as spots of color in juxtaposition, all the knowledge and much of the technique of the impressionist, or the pointillist school of painting, was immediately applicable to concrete. This school places color spots, side by side with one another, in such a manner that they blend in the air to hues of even value and chroma. The wonderful clarity of color, the vibrant quality of surface, which distinguishes the paintings of this school are found also in concrete. Colors for which no aggregate has been found yet, purples, blue-greens and yellow-reds are successfully made; simple colors such as blue, green and yellow are blended in various ways, producing remarkably beautiful compound hues which may be added to the number of the component colors as follows: blue, blue-green, green-blue, green, green-yellow, yellow-green, yellow. This wonderfully increases the number of colors available from a smaller number of good aggregates. The enrichment of our palette is of inestimable worth. Such a multiplicity of hues is a property of architectural concrete, unequalled by any other masonry material, in fact is rivaled only by the great medium of the mosaicist, which however

Fig. 106. GRAUMAN'S METROPOLITAN THEATRE, LOS ANGELES
Mural, "Princess of the Golden Kingdom" painted directly on the concrete.
WILLIAM LEE WOOLLETT, ARCHITECT.

Fig. 107. GRAUMAN'S METROPOLITAN THEATRE
Reinforced concrete cantilever trusses; the panels are filled with concrete tracery.
WILLIAM LEE WOOLLETT, ARCHITECT.

Fig. 108. GRAUMAN'S METROPOLITAN THEATRE
The low relief concrete ornaments were cast in place. The totem pole is made of black leaded glass.

Fig. 109. GRAUMAN'S METROPOLITAN THEATRE

The concrete was given a metallic appearance by applying layers of paint, Dutch metal, gold-leaf and varnish and polished by rubbing with pumice. The panne velvet hanging on the wall was designed by the architect.

Fig. 110. SHELL-HALL, CURE PAVILION, WIESBADEN, GERMANY
Pebble-stucco over ferro-concrete columns; the capitals are decorated with shells set in cement.

is not adaptable to form in three dimensions. One must not imagine that the blending of colors is as easy to do as to suggest. Success can be assured only by the highest technical skill in the use of aggregates, by colors which are sufficiently pure. Any red and any blue will not produce an acceptable purple, and only hues which have been properly balanced, as to value and chroma, will produce satisfying results.

Concrete is so wonderfully responsive that it has wound a spell around me and around the men in my studio. When the work is taken from the moulds each morning and the colors are exposed, there is something so spectacular, so magical about it, that our enthusiasm never abates. Many of the men have fallen under the spell to such an extent that they object to working with other materials. Architects and artists who have used concrete feel this attraction just as much as these craftsmen. Its use is being rapidly extended. The limit of its possibilities is not yet in sight and will not be, until a completed knowledge of its constituent materials and the phenomena attendant on them permits us to take full advantage of the properties of concrete. Even now,

Fig. 111a. HUDSON MOTOR PLANT, DETROIT
Entrance.
ALBERT KAHN AND ERNEST WILBY, ARCHITECTS.

in its present state of development, concrete is an architectural material without a rival. In its simplest, as well as its most ornate form, it is extremely interesting. To me, after its facility and permanence, which have to do with what we may perhaps call the mechanical properties of concrete, its quality of surface, which is of its aesthetic properties, is the most attractive. The many faceted particles of aggregate make an almost perfectly diffusing surface with great depth and vibration...... The light, reflected from this type of surface, casts a haze over it which is pleasing and which may be augmented by soft coloring in the aggregate. The diffusing surface acts in much the same way as a thin veil over colored silks. ...... The quality of light reflected from the walls of this church permeates the atmosphere as delicate perfume might. This is primarily what an architect may expect from concrete. We have designed surfaces to fill the most exacting requirements and to meet the greatest differences in scale, surfaces which lose their texture and resolve to uniform hue at twenty-five feet, surfaces which hold their text-

Courtesy *Arch. Record*
Fig. 111b. CALIFORNIA PETROLEUM SERVICE STATION,
WILSHIRE BOULEVARD, LOS ANGELES
Stuccoed concrete with applied red, black, and yellow Tunis tiles.
ROLAND E. COATE, ARCHITECT.

Fig. 112. DAWSON WAREHOUSE, STOCKTON, CAL.
Painted precast concrete frieze.

ure at five hundred feet. For these purposes, aggregates, measuring from less than one-quarter to more than one and one-half inches, were used . . . . . concrete did not once necessitate a modification of design. We were able to produce every form, color and texture which the architect required. Materials for this work were gathered from many places on this continent, from France, Italy and Africa. Materials so rare and hard that they have been beyond the reach of builders; quartz and onyx which are only used as jewels; ceramics colored by the oxides of rare metals and burned to melting heat. Particles of marble plated with gold, materials such as might fill the dreams of an Arabian night were moulded into the walls of this church. All this was made possible by concrete because in concrete these materials could be used in granular form; could be flowed into place with water, the great vehicle of creation; and could be bound firmly into a permanent solid by the magic of portland cement and Atlas Portland Cement was selected as the cement best suited to the work . . . it was a real test of the architectural value of concrete, a test with maximum requirements. And the suggestion is made that the performance of concrete under these conditions, is conclusive evidence of the ease, spontaneity, economy, perfection of form, color and textures, and the other marks of a great architectural medium which make it . . . . . .superior to comparable materials."

Oswald C. Hering suggests rolling colored sands into the surface of wet concrete where a smooth surface is desired[30]:

"A ceiling may receive this colored sand finish by plastering the bottom of the form with wet clay, to cover up the board marks. The upper surface of the clay is then sprinkled with the colored sand and the concrete poured in and well puddled and tamped. After the ceiling slab has hardened and the forms have been removed, the clay is washed off, leaving the sand adhering as a face to the concrete slab."

(30) "Concrete & Stucco Houses" p. 89.

Fig. 113. DAWSON WAREHOUSE, STOCKTON, CAL.
Window.
Glazed tile inserts and precast concrete ornament.

He further points out that a certain amount of variation in color is essential rather than a handicap to the interest and beauty of the wall.

Portland Cement Stucco.

Probably the most widely used material for the decoration of monolithic concrete structures is portland cement stucco (Fig. 90-94, 39). When it is used for interior decoration it is usually referred to as portland cement plaster. The range of the decorative uses of this plastic material is very great. In order to secure satisfactory stucco it is necessary to give much attention to creating an effective bond. If oil or grease is used on the form work it is apt to prevent effective bonding of the stucco which is later applied to the concrete surfaces. In all cases, dirt, grease, and form oil must be removed from concrete surfaces before either a single

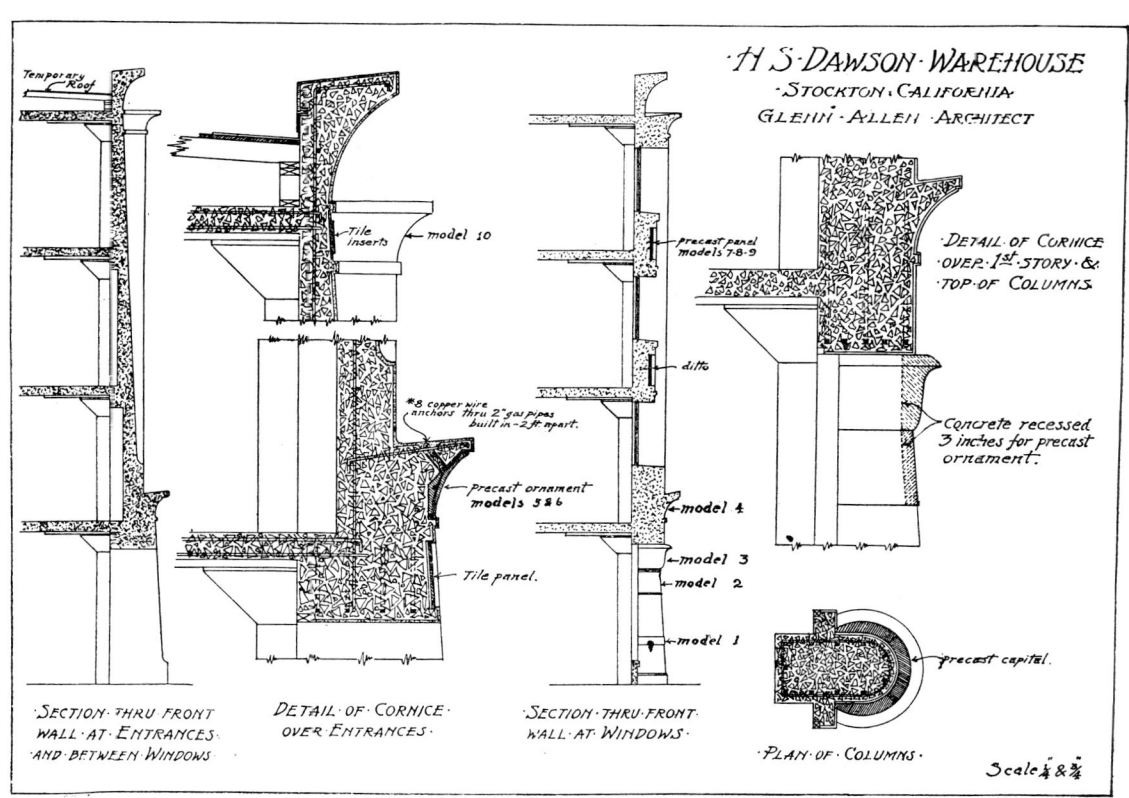

Courtesy *Concrete*
Fig. 114. DAWSON WAREHOUSE.
Section of Warehouse front showing how ornaments were secured in place.

Fig. 115. DAWSON FIREPROOF STORAGE BLDG., STOCKTON, CAL.
GLENN ALLEN, Architect.

Fig. 116. STUDIO FOR MME. C. O., SCULPTRESS, CITE SEURAT, PARIS
Diagonally placed glazed bricks create the textured panel. (1926).
A. & G. PERRET, ARCHITECTS.

Courtesy *Concrete*
Fig. 117. CONCRETE PANEL WITH COLORED GLASS INSERTS, YONKERS, N. Y.
J. HOOPER, BUILDER.

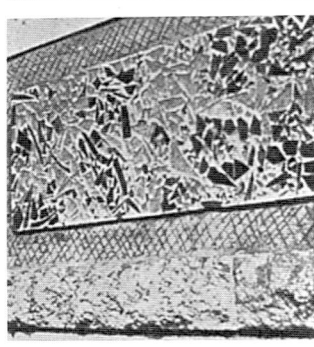

Courtesy *Concrete*
Fig. 118a. WALL FOUNTAIN WITH CHILDREN'S COLORED BEADS INSERTED IN CONCRETE

118b. CONCRETE PANEL WITH FINELY GROUND BLUE BOTTLE GLASS, YONKERS, N. Y.
H. V. WALSH, ARCHITECT.

dash coat stucco or one of the heavier stuccos is applied.

When two-coat or three-coat stuccos are employed, it is usually advisable to roughen the surface of the concrete before the application of the scratch coat. This may be done by removing the vertical forms before the concrete is too hard and then immediately brushing the surface with wire brushes. If the surface is allowed to harden too much the cost of roughening will increase greatly. A special patented liquid material which facilitates the removal of surface layers of cement mortar is now obtainable. This liquid material is spread on the interior surfaces of the concrete forms and here it retards the hardening of the cement mortar with

Fig. 119. PORTLAND CEMENT ASSOCIATION BLDG., CHICAGO, RECEPTION HALL
Art-marble lightstands and booth-dado; precast floor-tile.
HOLABIRD AND ROCHE, ARCHITECTS.

Fig. 120. PORTLAND CEMENT ASSOCIATION BLDG.
Elevator Lobby.
Painted concrete beams. Art-marble dado and floor tile.

Fig. 121. POLISHED CONCRETE DADO, R. R. STATION, KARLSRUHE, GERMANY

which it comes in contact. If the forms are stripped within definite time limits, the aggregate may be easily exposed by brushing the concrete surface immediately after the forms are removed. The surface is washed with water after brushing and allowed to cure before the application of the scratch coat. A good bond is, of course, at times not only dependent on the anchorage afforded by a roughened surface, but it is also dependent on the amount of suction that is obtained. It must be realized that concrete saturated with water will have practically no suction, whereas thoroughly dry concrete will probably have too much.

For a single dash coat a mixture made in the proportions of one cubic foot of portland cement to one and one-half cubic feet of sand has proved very satisfactory when just sufficient water has been added to make a grout of thick, creamy consistency. This mixture should be dashed on the concrete surface with considerable force by means of a stiff brush and allowed to harden without any trowelling whatsoever. Such a coating not only gives a more uniform color to a monolithic concrete structure but it also acts as a seal and diminishes the coarse appearance of the surface. It does not hide completely the texture of the monolithic concrete that lies beneath it (Fig. 95, 96).

The first coat of two-coat or three-coat stuccos should be a single dash coat such as has been described. The second and third coats may be trowelled on. Satisfactory mortars for these coats may be made of mixtures consisting of five sacks of portland cement, one fifty-pound sack of hydrated lime and sixteen cubic feet of sand; the finishing coat should have in addition the correct amount of mineral color pigments. Factory mixed Portland cement stuccos are recommended for the finishing coat to assure more uniform color. The first coats can also be placed by the cement-gun.

Scraffito.

"Hard scraffito" to quote from D. C. Allison[31] "is the result of applying successive layers or coats of different colored plaster and, after it has taken its set, the design is scratched thru in different parts to the color best suited to its representation; this technique employs as its means a combination of relief, texture, and color. The cement-gun can spread smooth surfaces of uniform thickness and the pneumatic hammer or tool can cut and surface hard granular material; these two machines in addition to the variety,

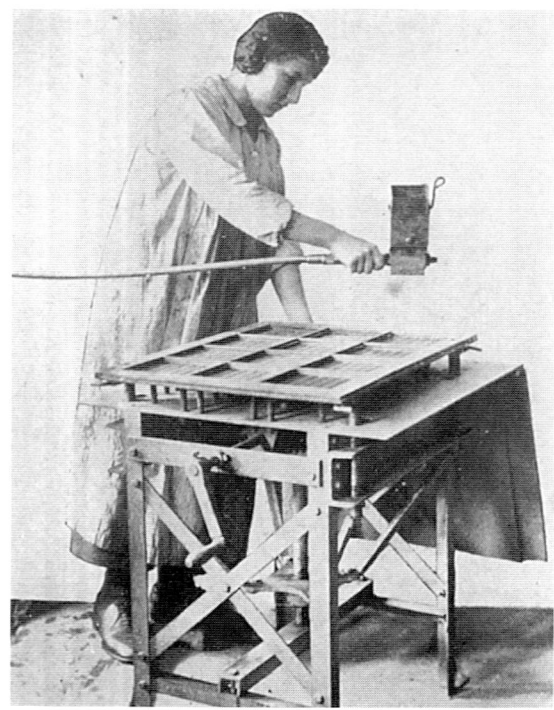

Fig. 122. A. WEITHALER'S SYSTEM OF GLAZING CONCRETE TILES WITH "GLASIN"
(Cold Process)

Fig. 123. CONCRETE RELIEF GLAZED WITH A. WEITHALER'S "GLASIN"

brilliancy, durability, and workability of the cement-mortars available to-day represent a big advance when contrasted with original methods of creating scraffito (Fig. 97)."

"Soft scraffito is the method of placing plaster on a wall in certain definite areas which constitute the design to be executed. Wet plasters, when laid side by side, offer the same opportunity of blending as do paints and intermediate hues can be used and so soft scraffito can be considered a kind of fresco. Stencils cut from light roofing materials or sheet metal, carefully studied for overlapping, are used to build up successive layers of different colored portland cement mortars into a design, either by hand or with the cement-gun; the stencils withstand the sandblast of the cement-gun for a considerable length of time. When the artisan is not a designer, the stencils afford a mechanical means of executing the will of the artist to a high degree."

---
(31) "Proceedings of the Amer. Concr. Inst.", 1926.

The facade of the Women's Athletic Club of Los Angeles, designed by D. C. Allison, was covered with scraffito in tones of buffs, browns, blues and greens.

Painting.

Stains, paints, enamels, and varnishes are mostly used for the interior. Mr. J. B. Mason makes the following helpful suggestions concerning the painting of concrete:[32]

"Color should be used ... to assist the ornamental schemes in their mission of emphasizing construction. Strong accents are placed at important points. For example, radial elements such as circular borders and interlaces may reinforce the upward thrust of a dome or rotunda .. Focal points marked with appropriate spots may help to correct the illusion that a beam, no matter how true it may be, seems to bow downward. Use of the beamstructure, supporting the floor above, as part of the decorative plan is both economical and sensible, since it allows omission of lathing, furring and plaster, and brings the beams into use artistically as well as structurally. It is some-

---
(32) "Architecture", Sept. 1927.

Fig. 124-125. "KERAMENT" COLD-GLAZED CONCRETE TILE AND SCULPTURE

times necessary to add a few false beams to complete a design, but as a rule this expense is considerably less than the cost of lath and plaster.... The concrete surface texture, in the most successful work, is made a part of the design. Geometrical forms, as a rule, are more appropriate to concrete than floral ones."

Preparation of the Surface

Sufficient time under the proper conditions of temperature and humidity must be allowed for drying or the moisture in the concrete will eventually injure the oil-paint coating. It must be realized, however, that concrete should not be permitted to dry until after a proper period of curing has ensued. The concrete must be kept damp during the first week or ten days after placing in order to assist the chemical reactions which give the concrete its strength and its durability. More elaborate methods in the preparation of concrete to receive decorative paints is required in the eastern part of the United States than in the West due to climatic and moisture conditions.

To neutralize the lime near the surface of the concrete, which would destroy the paint oils, Mr. Mason suggests three methods:

1) Painting the concrete with eight and one half pounds of zinc sulphate dissolved in a gallon of water and allowing 48 hours for the reaction.

2) Sponging the concrete with a seven-per-cent solution of Magnesium fluosilicate; this has not the waterproofing effect that some expect it to have.

3) Allowing the concrete to weather. Concrete that has weathered for a month or more, no longer contains any lime near the surface, provided it has been kept moist.

W. W. Clifford suggests one of the various excellent proprietary paints which are especially prepared to resist alkalinity and fill the pores. He

Fig. 126. KERAMENT COLD-GLAZED CONCRETE RELIEF

further states: "It is also necessary to exclude moisture permanently from a wall which is to be painted, as moisture working in behind a coat of paint will cause it to peel."

After neutralizing the surface-lime the pores of the concrete must be filled with raw linseed oil with or without pigments, or linseed oil containing red lead with sufficient drier, and if the concrete is fairly dense, turpentine or naphtha thinner should be added to get penetration. Spar varnish cut down one-half with turpentine, to which some aluminum-bronze powder has been added, also fills the pores. Concrete which is of an extreme porosity requires two priming coats.

Paints

The primed concrete retains all of the texture and is ready for staining with thinned linseed oil paints; the stained surface can serve as a background for stencils or murals. Paints of pure linseed oil ground with standard pigments give excellent results; tung oil is sometimes added as it toughens the paint film and gives increased resistance to moisture. Non-drying oils should not be used as they

Fig. 127. FOUNTAIN IN BRESLAU, GERMANY
Kerament Cold-glazed Concrete Tiles.

cause disintegration. In the final coat it is advisable to use 90 per cent raw linseed oil, 10 per cent turpentine and one pint of good copal varnish to each gallon of paint. Various combinations of white lead, lithopone, zinc oxide, titanium oxide and inert pigments mixed with either linseed oil or heavy-bodied enamel liquids have given excellent service. Tinting pigments should be alkali-proof and, of course, light-fast. Ready-mixed paints made by reputable manufacturers especially for the painting of concrete surfaces can generally be relied upon to give satisfactory results.

Some German builders apply mineral or casein paints immediately on a concrete of fine-grain surface produced by a smooth mold. They consider glue colors preferable to oil-paints for stuccoed concrete. For in-

Courtesy *Concrete*
Fig. 128-129. "METALLIZED" CONCRETE. (HOLLAND).

terior work Mr. Twelvetrees recommends waterpaints which can be used when the concrete is not perfectly dry.

Protective Coats

To protect the painted surface, thin, clear varnishes or shellacs are often applied. By their use a glazed surface may be produced or an antique effect may be given to an entire color scheme. An additional protection that assists greatly in the cleaning of the surface is a coating of starch. When such a coating is removed by washing, any accumulation of soot or dirt on the surface is also removed. Afterwards, the surface may be recovered with a fresh coat of starch. The addition of such protective coatings, of course, increases the economy that is obtained by the staining and painting of decorations directly on concrete.

Examples

Fig. 102-104 show an admirable example of painted concrete on which the architects W. Gehron, S. F. Ross and M. W. Alley make the following comment:

"The modern trend in architecture leads to the utilization of existing materials to obtain a desired result with the least expenditure to the owner. In the engineering of the ceiling the structural members were placed to conform to the given design, and a number of false beams were added as required. The mixture used for this ceiling was cinder concrete as a matter of economy, which in this instance was a mistake, as it proved to make the problem of painting much more serious on account of the poor quality of the present day cinders obtainable, and the chemical reaction of the sulphur content on the base coat of paint. A large sized section of the ceiling was cast as a model, using the same mixture as the ceiling to be decorated, on which numerous cement

Fig. 130. CONCRETE FOUNTAIN, DUSSELDORF, GERMANY
(Exposition 1902).
PROF. JANSSEN, SCULPTOR.

Courtesy *Concrete*
Fig. 131. CHECKERBOARD DESIGN, FAIRMONT BRIDGE OVER MONONGAHELA RIVER, W. VA.
Made by 12 in. square, 1-2 in. thick inserts in form.

paints and base coatings were applied; all of these proved unsatisfactory". The architects finally turned to a chemist, Mr. Nathan C. Johnson who devised a base containing cellulose to which was added a certain amount of color to permit visual inspection, insuring a thorough coating of all surfaces; after the coating had been applied it was found an actual penetration had been attained of from $\frac{3}{8}''$ to $\frac{1}{2}''$. Mr. Johnson points out that in using cellulose solutions as a primer it is necessary so to choose the solvent that it will have the ability to penetrate into the concrete sufficiently to insure full adherence, and further be sufficiently low in volatility to permit its application in a commercial and effective manner.

"As an added precaution a second base coat was used, and it was found that the rough surfaces of the concrete and form marks were to a great extent obliterated or softened down by the original coatings, which left a remarkably good textured surface for the decorative painting, for which we used an especially prepared lead and oil paint with various pigments. Subject matter pertaining to the early Jewish history was depicted on the various beams. The costumes and colors were authentic and taken direct from rare ancient books and manuscripts. The development of the Jewish Temples and Synagogues, starting with the Tabernacle in the Wilderness, were depicted in chronological order on the two large longitudinal beams. In the case of the figure decorations the outlines were for the most part ponced on or drawn direct and filled in by hand, while stencils were used for the smaller beam decorations. The coloring was extremely brilliant and after being completed the entire ceiling was glazed down to the required tone. Such a form of decoration should be subject to less damage which often results where plaster is not properly bonded or may become loosened by dampness or for other numerous reasons."

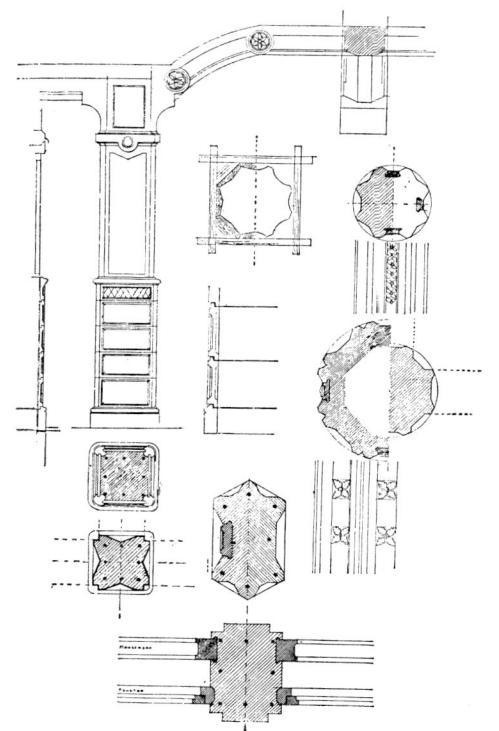

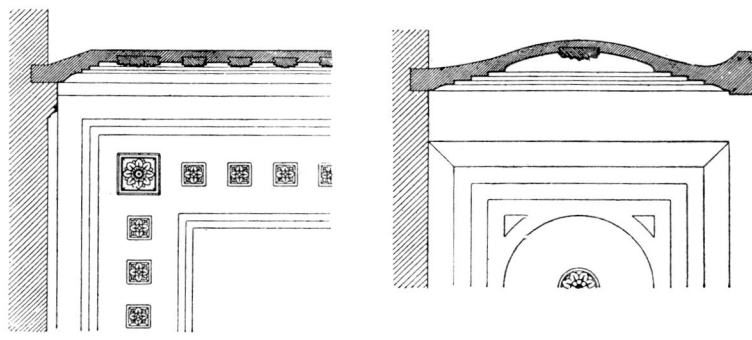

Fig. 132-133. DECORATION OF FERRO-CONCRETE MEMBERS BY PRECAST ORNAMENTS. (Austria)
Von Wielemans, Architect.

Fig. 105 shows the design of the same architects for the Cadet Mess Hall, West Point; a solution of aluminum was used on the dense stone concrete to secure a porous surface; cellulose acetate paint was then applied.

In the Honolulu Y. W. C. A. and in the University Club, Los Angeles (Fig. 96) the concrete floors were stained and waxed. For the latter building its architect, D. C. Allison, gives the following description:

"The surface was lined off in squares of about 14 in. which were stained by two or three brush applications of a thin metal hardener, which carried the color into the surface from a sixteenth to an eighth of an inch; any color in the gamut of browns, reds, greens and buffs was possible to obtain and each square being treated individually gave absolute control of the variation in color. These floors are waxed and polished from time to time and after five years of wear have taken on a patine, depth and richness of color that are surprising. The transformation from an ordinary gray cement floor to one similar to rich old tiling suitable to the use of oriental rugs is most gratifying with a low appropriation."

Mr. W. L. Woollett, the architect of Grauman's Metropolitan theatre in Los Angeles (Fig. 106-109) describes his methods as follows:[33]

"Nearly all of the concrete of the trusses over the auditorium and the beams and cantilevers on the mezzanine floor, and the proscenium opening with its flanking columns, are in concrete left in the original color, except for limited areas which were colored with pigments or metallic glazes. An all-over coat of very thin asphaltum will suffice to remove the appearance of newness. So little color or metal was applied to these surfaces that they count in color as raw concrete. These surfaces are far from the eye and were left purposely grey so that the color effects in the lighting scheme could be made effective incidentally. There is no more effective medium for electrical display than the neutral tones to be obtained in raw concrete. For nearby surfaces ordinary lead and oil paint is a good medium, dragging over it either color in oil or color mixed with benzine or turpentine. If the surfaces are sufficiently far from the floor areas, to be free from the accidents occasioned by cleaning and dusting, water color will serve the pur-

(33) "Concrete", Aug. 1923.

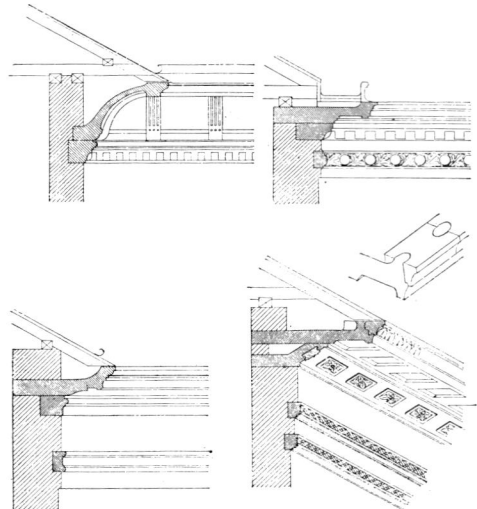

Fig. 134. DECORATION OF CONCRETE CORNICES WITH INSERTED PRECAST ORNAMENTS (1908)
Von Wielemans, Architect.

Fig. 135. FERRO-CONCRETE CEILING IN THE JURY-COURTROOM, SALZBURG, AUSTRIA
V. Wielemans, Architect.

pose. It is also practical to coat the concrete with a good varnish and pounce the color into the varnish. It is commonly known that the chemical reactions which take place in freshly formed plaster or concrete affect pigments which are applied to these surfaces and it has been my custom to use only pure color mixed with lead or zinc, so far as possible, looking forward with expectancy to the action of the chemicals upon the pigments to produce an uneven and mottled appearance. One can modify the effects of the chemical reactions by using sizing, depending upon the amount and character of the reactions desired.

"It is sometimes necessary to paint the walls in a fuller color and with stronger contrasts than one would if one were not expecting the fading due to the action of the chemicals in the wall. It is also possible to bring out designs in the wall through the use of resistants . . . . . In regard to large areas which must be colored in a satisfactory manner the rough surfaces of the concrete are much cheaper to handle than ordinary smooth plaster surfaces, produced by the ordinary method. Decorators spend a great deal of time and money in producing what is called "Tiffanying" on plain wall surfaces. This process produces variation of color . . . . : a body tone with one or two superimposed tones in the form of glaze perhaps, which were pounced with a dry brush or cloth, gives a cloudy or stippled effect . . . On concrete surfaces this is a simpler operation than on a plain plaster surface. An interesting effect is obtained on concrete by first applying a body tone and after it's dry, dragging the surface in such a manner that only the projecting areas are covered with a second coat—or the process may be reversed by painting on the second coat and rubbing off the high lights. The projecting surfaces may be dragged in another direction with still another color, or the second dragging or first dragging may be done with shellacs, varnishes, beeswax or paraffine, according to the effects desired. In addition to the varying tones so easily obtained on rough concrete, one also obtains the sense of permanence of the material, which adds to the value of any work of art. For concrete surfaces which are near the eye and which have been elaborately modeled, it is desirable to cover the entire surface with Dutch metal or aluminum leaf, or even with gold leaf, and afterwards apply color broken up in oil or mixed in turpentine,

Courtesy *"The Architectural Record"*
Fig. 136. VILLA CARLOTTA, HOLLYWOOD, CAL.
Arthur E. Harvey, Architect.

benzine or water, according to the effects desired. Sometimes we cover carved or modeled objects—such as the great lion in the foyer of the Metropolitan Theatre with alternate layers of Dutch metal and oil paint, also alternating varnish and shellac with Dutch metal and color, and sanding same to a glass finish, very much as one would finish a piano top. This process, if the concrete has been roughly cast, does not take away from the effect of the object being cast in concrete but does give beautiful surfaces which are friendly and effective."

Spraying.

Colors have been sprayed on concrete buildings with good results. The roofs of the exposition halls built for the Dresden Hygiene Exposition were covered with thick green color mixed with sand by means of a machine similar to the cement-gun. By "shooting" sands of various colors, effects like those achieved by Mr. Earley might be attained with a cement-gun.

Reinforcing Veins.

As several authors have remarked, the unity of reinforced concrete construction eliminates base, capital and the like parts as separate members. One part develops out of the other just as members of organic bodies do. According to John Ruskin, "Truth does not demand that the constructive core be laid bare; a protecting skin which with organic bodies covers the bones and muscles is justifiable likewise in buildings." Continuing this simile, veins of darker colored aggregate on the surface, indicating the hidden reinforcement, would enable

Courtesy *Celite Co.*
Fig. 137. ORNAMENT CAST WITH PLASTER MOLD IN PLACE, FIRST BAPTIST CHURCH, LOS ANGELES
Crushed rock aggregate with 3 lbs. Celite per bag of cement—1:2½:3½ mix.
ALLISON AND ALLISON, ARCHITECTS.

Courtesy *Concrete*
Fig. 138. CONCRETE FLOWER POT.

Fig. 139. CONCRETE FRIEZE, YORK RITE MASONIC TEMPLE, WICHITA, KANSAS
Eberson & Weaver, Architects.

the layman to apprehend the strength of ferro-concrete and the slender members would become more pleasing. Such veins would not merely follow the main outline of the reinforcing bars but would replace their angular shapes by curved ones, thus becoming esthetically valuable.[34] These veins would accentuate the contrast to stone and brick and show the unity of reinforced concrete. The best method for producing them would have to be ascertained by tests. Iron-particles which have been used to create rust-spots in imitation of shelly limestone might serve the purpose.

DECORATIVE INSERTS

Intarsia.

Concrete-intarsia is produced by fastening strips of wood, gypsum, or metal on the interior of the forms according to the lines of the ornament; when the concrete has set, the forms are removed and the grooves are filled with colored cement mortars.

Tile.

Colored tiles form a very effective surface decoration (Fig. 111, 141). The thinner tiles are usually placed before pouring—for ceilings directly on the forms, and for vertical surfaces glued to paper or to canvas or set in forms with clay. Tile is also laid in precast concrete slabs which are set into recesses created by the formwork. O. C. Hering states that when unglazed tiles are used their backs should be soaked in water and their face oiled as protection against efflorescence—an unnecessary precaution with glazed tile. A piece of felt placed on the face of the tile also protects from surface staining by wet concrete. The back of a tile should never be oiled as it would lose its porosity and power of adhesion.

The H. S. Dawson storage warehouses in Stockton, Cal. (Fig. 112-

---

(34) Flat reinforcing bars were bent into parabolas to follow the lines of maximum tensile stresses in the early days of reinforced concrete, according to W. W. Clifford ("The Architectural Forum", May 1922, p. 178).

Courtesy *Concrete*
Fig. 140. CONCRETE FLOWER BOX

Fig. 141. AZTEC HOTEL, MONROVIA
Lobby.
Ferro-concrete beams decorated with inserted tile.

Fig. 142. AZTEC HOTEL, MONROVIA, CAL.
R. B. STACY-JUDD, ARCHITECT.

Fig. 143. AZTEC HOTEL, MONROVIA
Entrance.

115) show a successful application of tile and paint concerning which "Concrete"[35] states the following:

"Stucco consisting of 1 part white cement (Atlas) and 2 parts Monterey white sand mixed to a consistency of a thick cream was whipped on to the building with brushes after a brown coat of cement mortar placed on the rough concrete was brought to a true surface. The panels around the doorway are in special handmade burned clay tile while the window panels and other ornaments are of precast concrete decorated in polychrome; all were set before stucco was applied to the building. The frieze is of dull glazed tiles; typically Egyptian brilliant colors were used. Fig. 114 shows how the precast ornaments were attached to the concrete structure. These ornaments were made in glue molds; the facing mix of 1 part Atlas white cement and 2 parts white Monterey sand was tamped in place with a 1:3 mix of ordinary cement and sand as backing and the whole reinforced with $\frac{1}{4}$-in. bars to give strength for handling. The polychrome decoration, consisting of specially prepared enamel

---

(35) July 1921 & Oct. 1922.

Fig. 144. AZTEC HOTEL, MONROVIA
Fireplace.
R. B. STACY-JUDD, ARCHITECT.

(ground and mixed in one-half turpentine and half spar varnish for a binder coat) was applied after the ornaments were in place. A second coat of 1 pint turpentine to each gallon of varnish completed the base on which the various colors mixed in spar varnish were painted; they have proven durable and resemble mat glazed terra cotta."

The architect, Mr. Glenn Allen, employed similar methods in decorating the Brueck and Huisholt Building in Stockton:

"The cast stone was colored with French ochre. No. 8 copper wires built in the concrete secured the surface ornaments placed in the recessed concrete. The large modillions under the cornice were secured in place by building in 1½" gas pipes extending up thru the cornice to the top at each modillion. Copper wires cast into the modillions were extended up to these pipes and twisted over short iron bars laid across the ends of the pipes; the stucco consisting of a brown coat was then applied, brought up to the true surface and all members and moldings formed. This brown coat was cut out before the final setting for the tile inserts in the columns, the tile set and the stucco finished with a coat of dash; the ornaments and tile were cleaned off before the cement had set hard by stiff wire brushes. The tile was cleaned down with diluted muriatic acid."

Glass.

The ornamental possibilities of inserted "glass-stones" are described in the third chapter (Fig. 108, 250-255). Concrete blocks with a glass pane front have been used in Germany since 1910. Mr. J. Hooper in Yonkers, N. Y. molded reinforced

Fig. 145a. MAYAN THEATRE, LOS ANGELES
Morgan, Walls & Clements, Architects.

panels one inch thick, incorporating colored glass pieces in the surface layer. The glass bits were quite flush and appeared like mosaic. These panels were set in place on the rough body with cement mortar (Fig. 117). Two other panels were decorated by plastering pure cement over the panel surface and throwing finely ground blue bottle glass at it. Children's colored beads set smoothly into the surface enhanced a small wall fountain in the same residence (Fig. 118).

GLOSSY CONCRETE
Polishing.

When the concrete surface-layer is sufficiently dense and the aggregate polishable, concrete can be polished like marble by hand with a stone or by power (Fig. 119, 120). The concrete to be polished must be lean enough to show a large percentage of aggregate when polished. In Germany feltballs dipped into tin-putty and *fluat* are used to polish concrete; repeated application of the filler and hardening is necessary. Carborundum grinding should precede machine polishing. The Karlsruhe station (Fig. 121) has a dado consisting of a concrete surface layer containing basalt-dust and fine gravel which was polished so that it resembles marble.

An unusual tone of gray may be obtained by rubbing graphite into the pores of the concrete and afterwards covering the surface with shellac and varnish, and subsequently polishing. By placing layers of paint, Dutch metal, gold-leaf and varnish

Fig. 145b. MAYAN THEATRE, LOS ANGELES
Detail of Facade.

on concrete and by rubbing and polishing with pumice, a unique metallic appearance can be obtained (Fig. 107, 109).

Glazing.

Concrete can be glazed by spraying on the surface finely ground cement mixed with a bituminous mass and certain chemicals. This method, termed "cold glazing", is practised by several firms in Germany.

A. Weithaler's Glasin provides a durable surface, as proven by slabs which are still good after having been exposed to the weather for sixteen years (Fig. 122, 123). When used on the exterior, Glasin-slabs must be made acid-proof. They are manufactured in various colors and their cost is about one third of the cost of burnt tile. Thru spraying different colors on top of each other very soft hues are achieved. The qualities of Glasin-products place them closer to polished marble than to artificial marbles made of lime and gypsum. Interior

102

Courtesy "Pacific Coast Architect"
Fig. 146. MAYAN THEATRE, LOS ANGELES
Main Foyer.

Courtesy "Pacific Coast Architect"
Fig. 147. MAYAN THEATRE, LOS ANGELES
Stairs to Balcony.

Courtesy *"Pacific Coast Architect"*
Fig. 148. MAYAN THEATRE, LOS ANGELES
Side Stage Detail.
MORGAN, WALLS AND CLEMENTS, ARCHITECTS.

Fig. 149. FERRO-CONCRETE CEILING IN THE HASLACH-STUTTGART SCHOOL GYMNASIUM, GERMANY
P. Bonatz, Architect.

walls can be glazed directly as a whole, if protected from cold and draft by closing the rooms in question for a few days. This direct glazing of a wall produces jointless, washable surfaces, which resemble porcelain and are cheap. Walls in schools and bathing establishments have been treated with this process.

The *Kerament* cold-glazing process, which has been patented in many countries, consists of spraying a mixture of the type described above on to the yet moist or freshly moistened surface by compressed air. Besides tiles and architectural sculpture (Fig. 124-127), roofstones have been glazed by the Kerament system. It can likewise be applied to freshly plastered walls. All color combinations are attainable. The Kerament-glaze is waterproof and tests made by the Dresden Technical University Laboratory prove that Kerament products also resist frost.

Metallized Concrete.

In Holland concrete surfaces have been metallized since 1919. The solution of metallic salts which imparts the coloring is applied with a hairbrush and penetrates to an appreciable depth into the surface of the material (Fig. 128, 129). It is claimed that this treatment renders concrete impervious to rain and damp. A range of about 35 colors are used, and if desired the solution can be prepared to produce any combination of colors or mottled effects. The solu-

Fig. 150-151. BELASCO THEATRE, HOLLYWOOD, CAL.
CONCRETE STONE ORNAMENT (See Page 122.)
Morgan, Walls & Clements, Architects.

tion leaves the colored surfaces dull, but this can be polished to any degree of brilliancy by treatment with wax, French polish, or any other common method.[36]

SCULPTURE

The economic advantages of the plastic quality of concrete have been well evaluated by John J. Earley:

"In this great industrial era, the twentieth century, mechanical duplication has greatly increased the output of industrial workers. The corresponding increase in their wages has been reflected in the Arts where there are no similar methods of duplication. The cost of individual pieces made by well-paid craftsmen has become so high that architectural decorations composed of such units are no longer within the usual economic limitations of building and therefore new methods are now necessary.... The solution of this problem is the development into an architectural medium of a plastic material possessing strength and permanence. Such a medium, because of its plastic nature, would require less labor to mold and shape than is necessary with a solid material....

"How short a period of time has elapsed since concrete has been treated as a plastic of great mobility and not a granular solid to be pushed and tamped into place ....Craftsmen have always known that for the production of three dimensional forms plastic materials are the very best, and, because of the great facility of such materials, have used them wherever possible ... less force is required to mould and form a plastic than a solid. Now, therefore, because force whether mechanical or human may always be measured in terms of expense, plasticity endows a material with basically economic properties."[37]

As ferro-concrete unites the char-

(36) "Concrete", Dec. 1922.

(37) "Architectural Concrete".

Fig. 152-153. BELASCO THEATRE, HOLLYWOOD, CAL.

Fig. 154-155. BELASCO THEATRE, HOLLYWOOD

acteristics of stone and steel it encourages the sculptured representation of objects which have both fleshy and delicate parts. Leaves and petals, for instance, will consist principally of concrete with a little reinforcing, whereas stem and stamens can be made of wire coated with cement mortar (Fig. 1). The Corinthian capitals of the church of St. Mary Immaculate of Lourdes, Newton Upper Falls, Mass., have reinforcing wires fastened to the bell projecting into each leaf; thus the capital could be deeply undercut in spite of the climate which would have prohibited execution in natural stone. The figures in the tympanon were likewise made of concrete. The monumental fountain at the 1902 Exposition in Dusseldorf, Germany, sculptured by Professor Janssen, contains reinforcing bars in the projecting parts (Fig. 130). In Stuttgart, Germany, the sandstone figures on the attic of the Royal Palace were endangering pedestrians as parts of the sculpture broke off. These figures were replaced by concrete sculpture which received reinforcing bars in the extremities and in this way were made safe.

Cast Sculpture.

Patterns of many varieties may be produced in concrete by simply nailing wooden blocks on the flat form surfaces (Fig. 95, 131). The planes of the ordinary formwork may be easily modified to produce in the finished structure either raised or sunken lines or panels. Reveals and projections that are appropriate to the design can be readily formed so as to cast shadows that will add much to

Fig. 156-157. BELASCO THEATRE, HOLLYWOOD
Morgan, Walls & Clements, Architects.

Fig. 158-159. MUSIC BOX THEATRE, HOLLYWOOD
Morgan, Walls & Clements, Architects.

Fig. 160-162. MUSIC BOX THEATRE, HOLLYWOOD, CAL.

Fig. 163. RESIDENCE OF J. J. SOWDEN, LOS ANGELES
LLOYD WRIGHT, ARCHITECT.
"I have found in this house, built by my son, a treatment of the block that preserves the plastic properties of concrete as a material."—FRANK LLOYD WRIGHT.

the effect of the architectural composition (Fig. 16, 48).

W. W. Clifford describes the making of molds in detail in an illustrated article in "The Architectural Forum"[38]:

"Wood forms for cornices and moldings are either made of narrow longitudinal lagging on templates cut to the cornice outline or consist of solid pieces. Lagging has to be dressed by hand and hence is expensive; it is only used when solid forms cannot be combined satisfactorily. The fine, closely spaced joint lines of the lagged form will be much less objectionable than a single larger joint. When solid pieces are used joints should only be made at angles.

Forms for moldings or other ornamental work should be left as long as possible. The harder the concrete the less likely are the corners to chip.

Von Wielemans, an Austrian archi-

---
(38) Feb. 1922, p. 67.

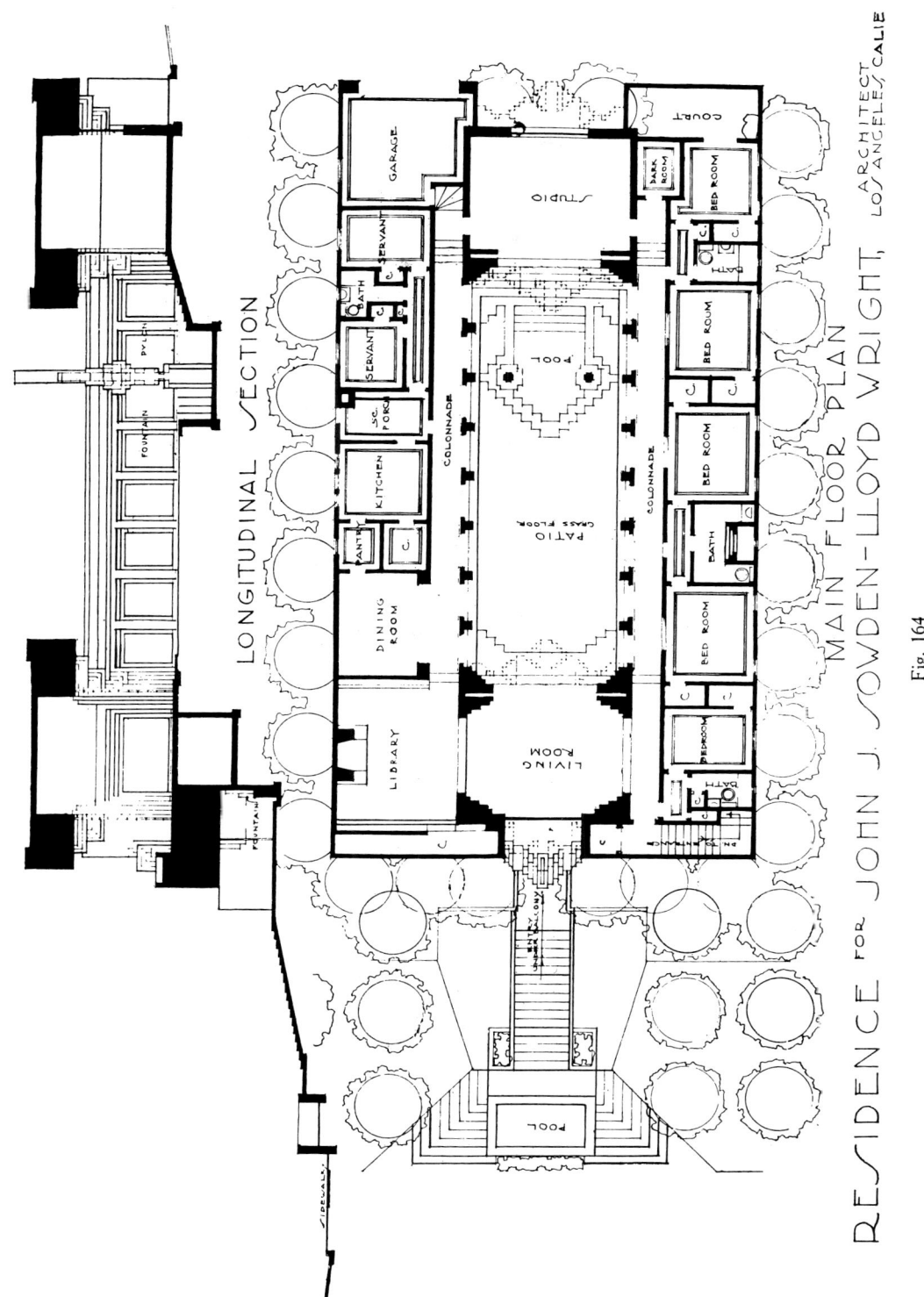

Fig. 164

RESIDENCE FOR JOHN J. SOWDEN—LLOYD WRIGHT, ARCHITECT, LOS ANGELES, CALIF.
MAIN FLOOR PLAN

Fig. 165. RESIDENCE OF J. J. SOWDEN, LOS ANGELES
Front Facade.
Precast concrete surface-slabs.

Fig. 166-167. RESIDENCE OF J. J. SOWDEN, LOS ANGELES
Pylons and Studio seen from the Patio.

Fig. 169. RESIDENCE OF J. J. SOWDEN, LOS ANGELES
Fireplace.
Precast concrete slabs.
LLOYD WRIGHT, ARCHITECT.

Fig. 168. RESIDENCE OF J. J. SOWDEN, LOS ANGELES
Pylon and colonnade seen from the patio.

Fig. 171. THE INNESS HOUSE, LOS ANGELES
Fifth Textile-block-slab house.

"Yet gradually the law of gravitation has its way, even with the profession: natural tendency in even so humble a thing as a building material will gradually but eventually force the architect's hand and overcome . . . 'professional resistance' . . . compounded of ignorance, animal fear and self-interest."*

FRANK LLOYD WRIGHT, ARCHITECT.

*This and the following sentences are quoted from Mr. Frank Lloyd Wright's article, "The Meaning of Materials—Concrete," in "The Architectural Record," Aug. 1928.

117

Fig. 172. THE INNESS HOUSE, LOS ANGELES

Block units made of decayed granite aggregate and white cement. Double wall shells. Perforated outer wall-shell in upper wall.
"It will faithfully hang as a slab, stand delicately perforated like a Persian faience screen or lie low and heavily in mass upon the ground."

FRANK LLOYD WRIGHT, ARCHITECT.

Fig. 170. THE INNESS HOUSE, LOS ANGELES
Retaining wall 32' high, single shell; each block 16"x16".
"Concrete? Is it Stone? Yes and No. Is it Plaster? Yes and No. Is it Brick or Tile? Yes and No. Is it Cast Iron? Yes and No."
FRANK LLOYD WRIGHT, ARCHITECT.

tect, inserted precast concrete ornaments which served as part of the mold as well as for decoration; they were used for columns, beams and domes (Fig. 76, 132-135).

When the assistance of skilled mold makers is available, plaster molds or glue molds of considerable detail and refinement may be made and subsequently erected together with the wooden forms.[39] Only one panel need be modeled and the plaster of Paris molds, built right in with the structural forms, will permit the casting of the ornamentation many times over. From one glue mold only five to eight casts can be made whereas the life of a plaster mold is practically un-

---

(39) R. C. Davison's "Concrete Pottery and Garden Furniture" contains an accurate description of this technique.

Fig. 173-74. CONCRETE STATUE AND TOTEMPOLE, MIDWAY DANCE GARDENS, CHICAGO

"It is a willing material while fresh, fragile when still young, stubborn when old, lacking always in tensile strength."

FRANK LLOYD WRIGHT, ARCHITECT.

limited. Glue molds are used when the design has heavy, undercut relief as they can be made in fewer pieces and can be more easily removed which decreases the danger of injuring the delicate parts of the green concrete. Care must of course be taken to see that the plaster molds are properly and securely fastened in place. Those faces of the plaster molds which give the form of the ornament should be protected from being spattered with concrete. If the mold is small this may be done by hanging a piece of canvas over it until the concrete is placed up to the level of the top of the mold, and then withdrawing the canvas and allowing the concrete to flow in. If the concrete is spattered on the surface of a plaster mold and permitted to dry somewhat before the concrete of the ornament is placed, it is apt to produce scars on the finished work. The concrete should be spaded from the middle of the form towards the face of the mold in order to assure proper filling and to eliminate air pockets. Sand moulds are often used and with them a pleasing texture can be economically obtained.

Surface decoration by low relief such as shown in Fig. 139, 140 can be more cheaply executed in concrete than in stone.

The Aztec Hotel in Monrovia, Cal. (Fig. 141-144) is noteworthy for the wealth of sculptured surface decoration which shows the possibilities of concrete for adopting any form. The bizarre Aztec forms may

Fig. 175-177. THE MONUMENT OF TIME, CHICAGO
Exposed Aggregates
Lorado Taft, Sculptor.
John J. Earley, Architectural Sculptor.

Fig. 179. "BLACK HAWK", EAGLES NEST BLUFF, ROCK RIVER, OREGON, ILL.
Lorado Taft, Sculptor.

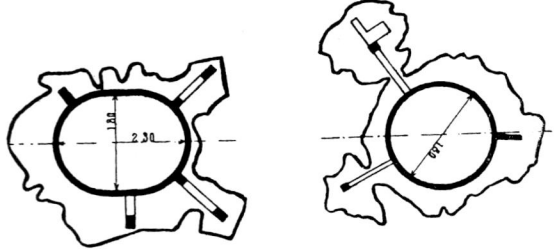

Fig. 178. FERRO-CONCRETE STATUE OF ST. JOSEPH, ESPALY near PUY, FRANCE
(The dimensions are in meters).

create the atmosphere desired and will serve the legitimate publicity interests of the establishment, but it would be deplorable if an "Aztec Movement" set in and style copyists were diverted from noble examples to the forms of a semi-barbaric people. It is hardly necessary to point out that these forms are in no way expressive of the Ferro-Concrete Style. The inlaid tile of the ceiling represents a technique which is easily executed in concrete and therefore can be recommended.

The architects Morgan, Walls & Clements developed an original technique for the cast concrete ornament of the Mayan Theatre, Los Angeles, (Fig. 145-149):

"The ornament was modelled in clay from which glue or plaster molds were made. If a light color was desired white cement in the mixture for the face was used; the stones themselves were wet cast and not tamped. The material for the facing contained no coarse aggregate and was wet mixed but did not contain enough water to puddle; it was laid in the mold with a trowel. By this means certain small voids were left in the surface of the stone which would have been filled if tamped. Care must be exercised in placing this moist face coating to distribute the voids as desired. A wet mix concrete filled the form as backing. Waterproofing was accomplished by the

Fig. 180. SCULPTURED CONCRETE PANEL.

use of integral compounds or by an external treatment of a clear waterproofing compound applied after all decoration was completed. The stones were decorated by painting on a water mixture of Hydrotite paste and pure limeproof color after they had set. The antiquing process consisted of the application of varying glaze color coat."

Frank Lloyd Wright creates very interesting sculptured walls, "..... textile-block-slab construction in which the method of building wholly determines the form and style" (Fig. 163-172, 239, 240). He describes his method as follows:

"The block units 16"x16"x4" are manufactured on the ground in metal flasks provided with dye faces in any pattern desired; the aggregate is tamped into the forms by hand. When assembled a continuous core is formed around all blocks in which various size steel is placed dependent upon stresses to be met and wet grout of 1:2 concrete poured in, thus locking the blocks and forming a monolithic slab, reinforced 16 inches o.c. both ways with $\frac{1}{4}$ in. steel rods which have been placed in the joints previously; these joints are filled by the poured conrete and become invisible. Double walls are constructed, tied together with steel clips, for air space between.. These buildings are practically mono-material structures, the cement block providing finished surfaced walls and floors. The possibilities for variation in design, either in color surface, texture, over all pattern or variation in block mass are endless. The blocks are constructed of a one to four or five mix of conrete, wet or dry as required for texture or stresses, with various colored aggregates used, the aggregate being the same throughout the block."

In some cases alternating blocks have a smooth surface. In LaJolla Mr. Wright had the walls poured in horizontal layers. Each layer was separated from the next by a V-groove formed by a wooden strip that was nailed to the mold. In one of his buildings the V-shaped joints on the interior are featured by dark mosaic stones set in cement mortar, and the concrete is not covered with plaster. Occasionally the ornamented slabs are pierced to admit light.[40] This sincere ornamental treatment of concrete units shows how ridiculous our ordinary "rockface" concrete blocks really are; petrified tapestries can now enclose our spaces and clothe our masses.

(40) Further possibilities of this kind are discussed in Chapter III.

Fig. 181. ST. LOUIS, VILLEMOMBLE, FRANCE
Paul Tournon, Architect.

The sculpture in the Midway Gardens, Chicago, (Fig. 173, 174) designed by Frank Lloyd Wright, shows a rugged simplicity which is very architectural and expresses characteristics of concrete, though the angular shapes are somewhat influenced by the "wood-centring style".

Lorado Taft's Monument of Time in the Chicago Midway (Fig. 175-177), cast in concrete by Mr. Earley, shows what can be attained by ferro-concrete plus craftsmanship. Two hundred thousand dollars were saved by executing this monument in concrete instead of using stone as first contemplated; Mr. Taft considers it superior in color and texture to chiselled marble.

"Concrete"[41] describes the casting as follows:

"A massive concrete base was constructed and upon it built up an inner falsework. The outlines of this falsework

(41) Dec. 1922.

POPE GREGOR VIII

Fig. 182-183. CONCRETE SCULPTURE ON THE TOWER OF ST. LOUIS, VILLEMOMBLE
M. SARREBEZOLLES, SCULPTOR.

were very carefully worked out to keep six inches away from the extreme outer surface as finished, in order that the finished group might have a total thickness of concrete of six inches..... Over the falsework was placed a metal fabric and on this plastered a weak porous base of portland cement and sand. It was made porous in order to take the excess water out of the structural mixture in placing and give the strongest possible result. Following these preliminary steps, steel was placed wherever required for reinforcement. In the meantime plaster piece molds were built up around the plaster structure executed by the sculptor..... They were set up on the concrete base around the falsework and the large group was cast in sections in place.... When the concrete was hard, the plaster piece molds were removed and the surface washed down with acid. The concrete was made with Potomac River gravel, a material with which Mr. Earley has done a great deal of work in Washington. It is in beautiful yellow, buff and brown colors, and the careful grading, with surface coloring values in mind, which Mr. Earley does at his Washington plant, gives the final surface in which the percentage of cement visible is at a minimum. The aggregate is of a size that passes through a ⅜ in. mesh. A great deal of Mr. Earley's technic lies in the grading and final proportions of the mix. So carefully have his formulas been worked out, that he has been able to do combination construction where concrete cast in place, concrete applied as stucco and concrete in precast units (in each case made with the same proportions of material) meet in one surface of the finished work, without leaving any difference in surface to mark the end of one part of the work and the beginning of the other."

Another pioneer in architectural concrete, Mr. W. L. Woollett writes concerning the sculpture shown in Fig. 107-109.

"There are various ways of reproduc-

Fig. 184-185. CONCRETE SCULPTURE, ST. LOUIS, VILLEMOMBLE

St. Vincent de Paul. *St. Bernard.*
Fig. 186-187. DETAILS OF CONCRETE TOWER, ST. LOUIS, VILLEMOMBLE

ing architectural forms in concrete. The method employed chiefly in the Metropolitan Theatre was to model the forms in clay and from these, plaster casts in reverse form were made. These plaster casts were installed in position before the structural work was poured, thus making the ornamental and structural portions of the building a unit. The small moldings and arrises are inclined to spall off if the forms are not removed with care. To overcome this tendency the plaster forms are shellacked—a very thin coat, so that the moisture in the concrete may be absorbed by the plaster mold. The absorption of moisture from the concrete prevents a good set of the cement in a very limited film on the surface of the concrete. When the forms are removed from such a mold there is a thin film of powder on the surface of the concrete form, composed of unset cement and sand. In order to liberate the more elaborate architectural ornaments from the plaster forms it is wise to employ a skilled mechanic—a carver by preference—one who may be familiar with the contour of the architectural design, in order to save it from undue mutilation. Great care should be exercised in not making the reverse forms too heavy. The heavier the form, the more difficulty of course in cutting it away from the finished product. The thinner forms make it possible for the mechanic to follow the design."

Chiseled Sculpture.

Ornaments can be cut out of concrete with a strong knife blade 12 hours after the concrete has been poured.

Concrete is more easily sculptured than the harder stones if the work is done from eight to fourteen days after casting (Fig. 180). Compressed air tools are also employed to sculpture hardened concrete. A two-inch surface layer of fine grained aggregate enables the sculptor to chisel out

St. Francis of Assisi.   *St. Benoit.*

Fig. 188-189. HEADS CHISELLED IN CONCRETE CHURCH OF ST. LOUIS VILLEMOMBLE, FRANCE

fine profiles. According to an article in "The Scientific American"[42], Italian sculptors add vinegar to concrete to slow its hardening as this retardation gives more time in which to complete the chiseling. The work of a sculptor can be reduced to a minimum by an ingenuous form builder capable of blocking out or of bounding a statue by the flat planes which are already his stock in trade.

M. Sarrebezolles, a French artist, has shown the possibilities of concrete sculpturing on the tower of the St. Louis church in Villemomble, near Paris. The photographs (Fig. 181-192) show the architectural value of his work which was wrought in concrete of the same mixture as that of the tower proper, except that the large gravel was omitted. Concrete which has been cast in wooden forms could be chiseled the following day. Twenty figures in all were arranged in two tiers, each more than twenty-one feet high. There are eight small figures and a number of smaller heads to enliven the mass. The whole stands as a symbol of the church, with the great pope Gregory as the central figure. That M. Sarrebezolles had to do as many as four heads in three hours gives an idea of the required speed. The great undertaking was finished in three months; the rapidity with which this work was done indicates the reasonable cost. Thus ferro-concrete enables the architect to return to the Gothic tradition of making the building a petrified motion picture that grips the imagination of the spectator.

(42) 1911, p. 279.

St. Augustin.

St. Paul.

Fig. 190-192. CONCRETE SCULPTURE, ST. LOUIS, VILLEMOMBLE
M. Sarrebezolles, Sculptor.

St. Peter.

Possibilities.

In addition to the above listed methods of sculpturing concrete D. C. Allison suggests the following techniques:[43]

Etching floors with acid as employed in the Cathedral of Siena: "It is believed that these white marble slabs were covered with wax, into the soft surface of which the lines were drawn with a metal tool, the real incision having been made by the use of acid held in place by sand or sawdust. Concrete floors would lend themselves to the process as readily as slabs of marble."

Etching a freshly finished concrete floor or wall by scratching the design into the yielding surface with a suitable metal tool: "The counterpart of the dry-point burr is present in this process and, as in the etching, must be partly or wholly removed after the surface is hard set. Fillers

(43) "Proceedings of the Amer. Conc. Inst.", 1926.

Fig. 93. AL MALAIKAH TEMPLE—CIVIC AUDITORIUM
Concrete tracery balustrades. Painted concrete arches.

composed of earthy pigment and wax or cement may be rubbed into the lines if the work is intended to carry any distance or if, for sake of cleanliness, a smooth surface is desired."

Modelling ornament into a plastic material on the inner face of wood panel forms: "Present needs are for a material easily worked like clay, but unlike clay it must not dry, crack, shrink nor loosen itself from the wood form; it should be strong and yet capable of being dissolved or softened in order to remove it from undercuttings. Experiments have been made with mixtures of clay, sand and glycerine; with sawdust and glue; with sand, water and flour ... The whole process is the reverse of what is to be obtained in the finished work but one soon becomes accustomed to think in terms of the matrix."

High relief, built in place, by means of the cement-gun: "Steel armatures are readily and economically covered by this means and the degree of finish to which the work is to be carried is limited only by the skill of the director of the nozzle. While a tooled finish of such work is possible and while a toolable mixture may be deposited by means of the cement-gun, yet it is doubtful whether the modeler who has gained a degree of familiarity with the nozzle will ever consent to any such subsequent finish of his work. There is a joy in the building of masses and of shaving them into planes while quite plastic which increases as familiarity with the tool progresses, and it is highly possible that through this feeling, a freedom of expression may arise sufficient to develop a new style of architectural sculpture, bearing a similar relation to our present formal modeling that a rough watercolor sketch bears to a finished painting of fifty years ago.

"While our friends of the 6H pencil have made possible, through the use of this delightfully flexible material, concrete, the support and construction of any building shapes that can be devised, it is now up to the addicts of the charcoal and 4B to realize that this is a material of the greatest architectural possibilities and one in the use of which, as they study its intelligent application, they can again build structures that will stand on their own feet architecturally, devoid of the sham and inconsistencies of much of our recently past work."

It has been shown that a wide range of techniques can be employed to decorate concrete surfaces. Hence the often encountered opinion that concrete surfaces lack architectural qualities is not justified. On the contrary, "liquid stone" presents so many possibilities, that it is hard to choose among them.

Courtesy *Concrete*
SEAT AND SHELTER, KAHLER PARK, SAN ANTONIO, TEXAS
The "straw-thatched" roof is also of concrete.
D. RODRIGUEZ, SCULPTOR.

Courtesy Western Architect.
FIRST CONGREGATIONAL CHURCH, OAKLAND, CAL.
Cement plaster on monolithic concrete walls—Doorway ornamentation pre-cast.
JOHN G. HOWARD AND ASSOCIATES, ARCHITECTS.

CHAPTER III
CONCRETE TRACERY

Fig. 193. RESIDENCE OF MR. AST, VIENNA—DOBLING
Prof. Josef Hoffman, Architect.

"Imagine a city iridescent by day, luminous by night, imperishable! Buildings—shimmering fabrics—woven of rich glass—glass all clear or part opaque and part clear—patterned in color . . . . . such a city would clean itself in the rain, would know no fire alarms—nor any glooms. To any extent the light could be reduced within the rooms by screens, a blind, or insertion of opaque glass. The heating problem would be no greater than with the rattling windows of the imitation masonry structure . . ."*

Frank Lloyd Wright.

In our days of rush and hurry, as people pass a building, they have no time to look at small ornaments and decorations. Only large spaces of even surface set off by outstanding features can attract the attention of the automobilist who flashes by, or the business-man who has time for only a glance. Mr. Ast's concrete residence (Fig. 193) partly demonstrates the resulting tendency in modern design. Yet the real solution of this problem will be brought about by the development of a new type of wall-aperture,—Concrete Tracery—which will be one of the main characteristics of a Ferro-Concrete Architecture.

(*) This and the following quotations are part of the article "The Meaning of Materials—Glass" by Frank Lloyd Wright which appeared in the July 1928 "Architectural Record".

133

Fig. 194-195. MUNICIPAL APARTMENT HOUSE, BUDAPEST, HUNGARY
Ferro-Concrete Balconies, balustrades and cornice.
Prof. I. Medgyaszay, Architect.

Transitional Tracery.

More than twenty years ago architect de Baudot employed concrete tracery of a simple type in the Paris church St. Jean de Montmartre (Fig. 61). Since then other architects have sensed that grilles are especially adapted to "liquid stone", as they can be created by simply placing inserts in the forms (Fig. 10, 57, 59-61, 107). In 1908 the Hungarian Medyaszay reported the utilization of these possibilities at the VIII. International Architect's Congress. His apartment houses in Budapest (Fig. 194-195, 374) and theatres in Veszprem and Sopron, Hungary (Fig. 196-199, 372, 373) show the beautiful effects to be achieved by traceried windows, balustrades and consoles. The Veszprem theatre is ventilated by a centralized system and with few exceptions the windows need not be opened. The reinforced concrete forms a uniform network for the surface which is to be glazed and closed hermetically. The concrete ribs have on the inner and outer side rebates which are varnished several times and which form a frame for the glass-panes; they are calked with oilputty which creates durable, cheap, easily formed and hermetically closed windows. The concrete web has no projections but very strongly curved Oriental-Hungarian outlines, the decorative effect of which is to enliven the big, smooth surfaces of the facade.

Wielemans' register grille (Fig. 76-B) depicts the metallic delicacy attainable in ferro-concrete by using glue molds. The Riverside Park in Hyderabad, India, boasts of still more intricate work executed in concrete with tiny trowels (Fig. 200). This bench was built on the site with rough reinforcement of wire mesh and iron rods to make the work sound in spite of its very light section. The finish is of white sand and has a fine texture,

Fig. 197a. THEATRE, VESZPREM, HUNGARY
Cornice, windows, pergolas and terraces are of ferro-concrete.
PROF. I. MEDGYASZAY, ARCHITECT.

which looks like marble. The Bahai Temple, Wilmette, Ill. (Fig. 208, 209) will have remarkable concrete tracery in the dome and windows; the slender sections of the ribs visible in the photograph of the window-mold are characteristic of reinforced concrete as they could not be executed in stone or terra cotta. Mr. Grosvenor Atterbury utilized concrete tracery as main means of embellishment for the fleche and windows of the Forest Hills Gardens Tower as well as for the garden-wall of the Parrish Art Museum at Southhampton, L. I. (Fig. 210-212). Professor C. L. Zaccagna created in his house in Mexico City (Fig. 213) concrete parapets with marble aggregates ornamented with tracery; the deep shadows make

Fig. 197b. THEATRE, VESZPREM, HUNGARY
Facade of Stage Wing.

Fig. 196. THEATRE, VESZPREM
Prof. I. Medgyaszay, Architect.

Fig. 198. THEATRE, SOPRON, HUNGARY
Concrete Tracery in the windows and consoles.
Prof. I. Medgyaszay, Architect.

it very effective. A further example is the roof balustrade shown in Fig. 214, which consists of flowers growing out of a reinforced horizontal stem. This tracery crowns the building most attractively especially with a blue sky seen thru its holes. The tracery of the balustrade shown in Fig. 215 has horizontal, vertical and diagonal reinforcing bars; the interstices were formed by hollow tin molds. The railing on the concrete residence in Belmont, Mass. (Fig. 216) is reinforced with aluminum wires, and requires no painting, as one of steel or wood does. The reinforced concrete balustrade combines safety and beauty, for unity of material is one of the essentials of good architecture: a wooden or iron railing on a concrete building will never be as perfect as a concrete one (Fig. 218-223). Simple geometric tracery occurs also on some of the concrete balustrades

Fig. 199. THEATRE, SOPRON, HUNGARY
The portico shafts are remnants of the former theatre. The main cornice, terrace, balconies, windows, chimneys and ventilator-caps are of ferro-concrete.
Prof. I. Medgyaszay, Architect.

on the 16th floor of the Shelton Hotel, New York designed by Mr. A. L. Harmon.

Figural Concrete Tracery.

The above examples are but a suggestion of the possibilities of figural concrete tracery. Every desired design and scene can be created by the silhouette effect of ferro-concrete backed by the curvilinear window-holes, thus giving by day a dark and at night a lit-up background to these stone-silhouettes. They will be very monumental as they need no bas-relief to be effective. The concrete framework of piers and girders which form the skeleton of the facade-wall (Fig. 224) will become veritable frames for pictures and ornaments wrought in concrete tracery. Gothic tracery bars were limited in thinness; reinforcement by aluminum wires will enable the creation of very thin rods in concrete tracery (Fig. 209). Gothic tracery served to strengthen the window-panes; in concrete tracery this will only be of secondary importance, since the new tracery will act as bracing for the bearing members. The reinforcement will tie concrete tracery and structural frame into *one*,—will make the entire wall a rigid unit pierced by holes.

The modern architect must take the psychology of our present day life into account. Hundreds of impressions such as those coming from billboards, electric signs and traffic-signals compete constantly for the attention of the pedestrian and automobilist; motion pictures accustom his nerves to strong effects and striking contrasts. The traditional ornaments and bas-relief sculpture cast too pale shadows to attract the man in the street, especially in our temperate zone and further north where there

Courtesy *Concrete*
Fig. 200. CONCRETE PARK-BENCH, HYDERABAD, INDIA

is so little sunshine. The sculptured frieze in Fig. 225 is quite subdued by the b l a c k window-holes and its contents remain unnoticed. The balustrade tracery of the Lodge shown in Fig. 226 is very effective whereas the adjoining low-relief frieze is indistinct. The Vienna bank-building, Fig. 227, is an example of how sculptured panels with deep shadows can lend interest to an otherwise plain facade. If executed as concrete tracery with the background pierced and closed with glass, they would be still more effective, especially at night with illuminated rooms behind. Ruskin's advice was never more needed than to-day:

"The architect's chief means of sublimity are definite shades...The power of architecture may be said to depend on the quantity (whether measured in space or intenseness) of its shadow ... And among the first habits that a young architect should learn, is that of thinking in shadow, not looking at a design in its miserable liny skeleton: but conceiving it as it will be when the dawn lights it, and the dusk leaves it; when its stones will be hot, and its crannies cool; when the lizards will bask on the one, and the birds build in the other. Let him design with the sense of cold and heat upon him; let him cut out the shadows, as men dig wells in unwatered plains; and lead along the lights as a founder does his hot metal; ....his paper lines and proportions are of no value; all that he has to do must be done by spaces of light and darkness; and his business is to see that the one is broad and bold enough not to be swallowed up by the twilight, and the other deep enough not to be dried like a shallow pool by a noonday sun."[44]

Concrete tracery will "cut out the shadows" and provide the high lights. In the temperate zone houses need so much window area to admit sufficient light that, in order to create some restful wall space, the architect is prompted to leave the remaining areas undecorated, and only use the window-hole with its deep shadow and clear outline as a motive. As M. A. Lurcat writes:

"To-day the practical details of a building, such as windows, with scale and

---
(44) "The Seven Lamps of Architecture", 3-VIII.

139

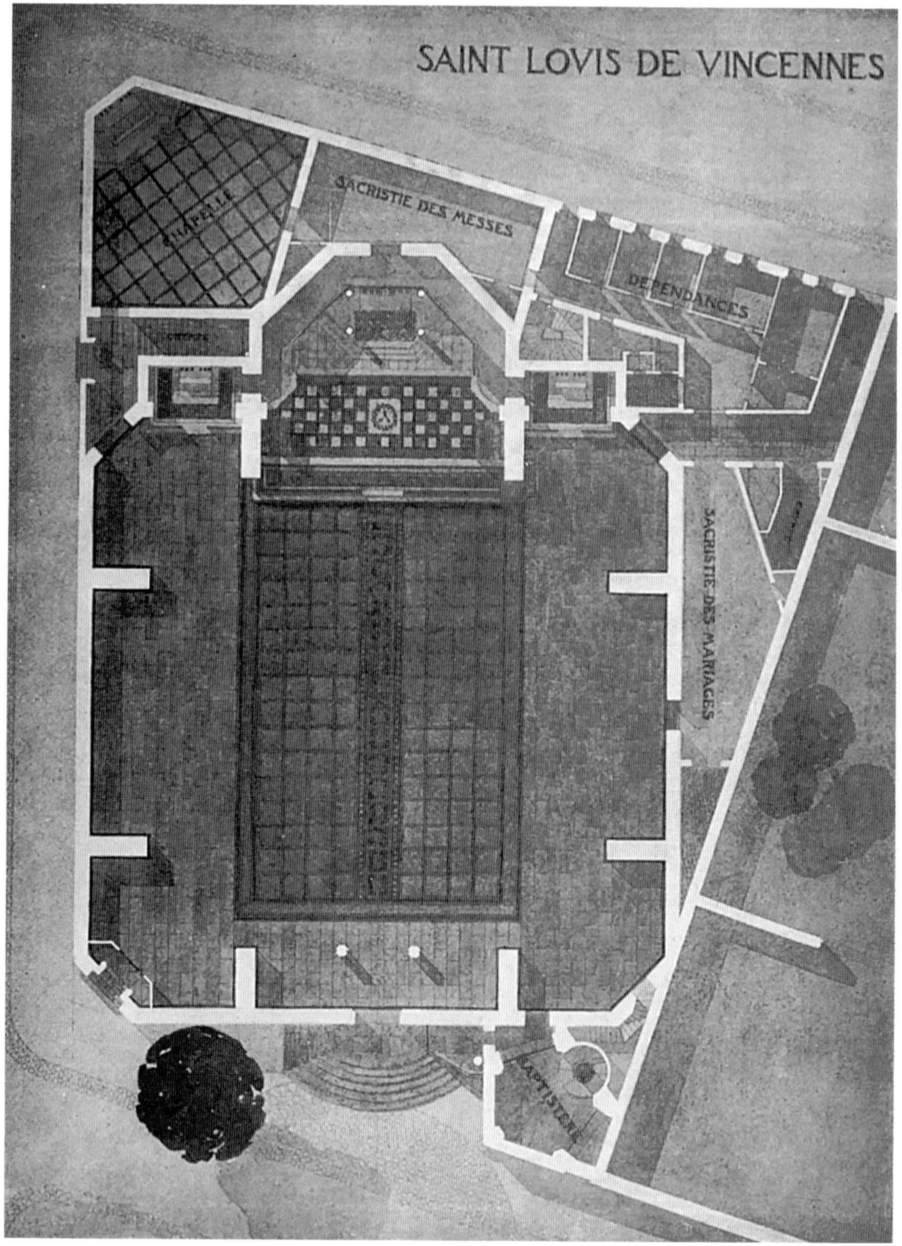

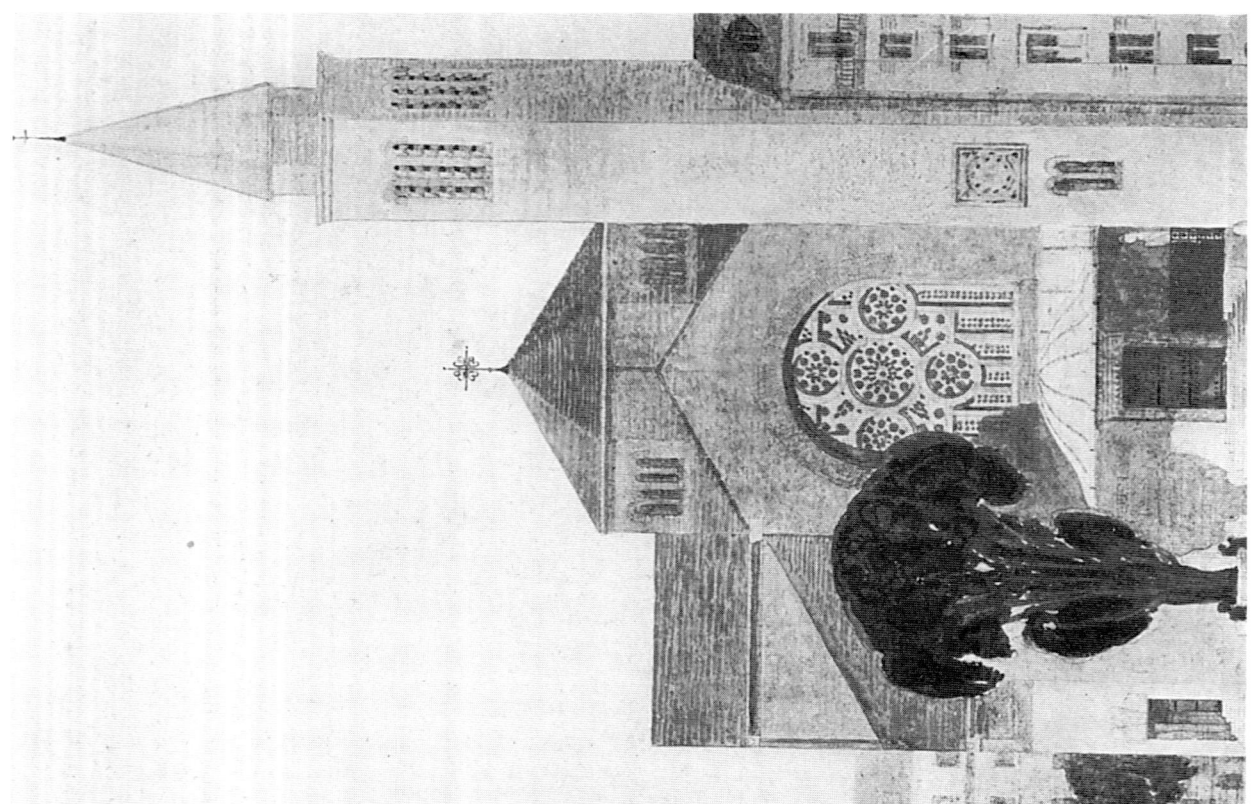

Courtesy *Journal of the American Inst. of Arch.*
Fig. 201-202. ST. LOUIS, VINCENNES, FRANCE
MM. Marrost & Droz, Architects.

Courtesy *Journal of the American Institute of Architects*
Fig. 203. ST. LOUIS, VINCENNES
Apse.
The frescoes were painted with water-colors on the fresh wall.

Fig. 204. ST. LOUIS, VINCENNES
Main Arches.
Ferro-concrete framework (Arches, tracery, floors) filled with masonry.

Fig. 206. ST. LOUIS, VINCENNES
Painted decorations and concrete tracery window.

Fig. 205. ST. LOUIS, VINCENNES
Polygonal lantern over the crossing.

Fig. 207. ST. LOUIS, VINCENNES
Choir.
MM. Marrast & Droz, Architects.

Fig. 208-209. BAHAI TEMPLE, WILMETTE, ILL.
(Concrete tracery will cover the dome and fill the windows of this temple when completed. The photograph at the right shows the model which will be used in casting the windows).
LOUIS BOURGEOIS, ARCHITECT.

proportion in the balance of solids and voids, have become its cardinal esthetic values."

The customary rows of rectangular windows are monotonous and hence the triangular openings visible in Fig. 228-230 and the octagonal ones in the covered bridge of the Dresden exposition (Fig. 231) were created. Mr. Piero Portaluppi of Milan has designed an apartment house with star-shaped windows which cover the whole facade as an all-over pattern (Fig. 232). In a second design (Fig. 233) he placed triangular balconies at regular intervals over the entire building for the same purpose. The curvilinear windows in the Garnisons Church at Ulm (Fig. 234, 235) are a step in the direction of this development. Variegated outlines of the apertures—concrete tracery—is the solution.

As concrete adopts any form into

145

Fig. 211. PANEL IN STATION SQUARE
Precast, inlaid blues, reds, greys and green.
GROSVENOR ATTERBURY, ARCHITECT.

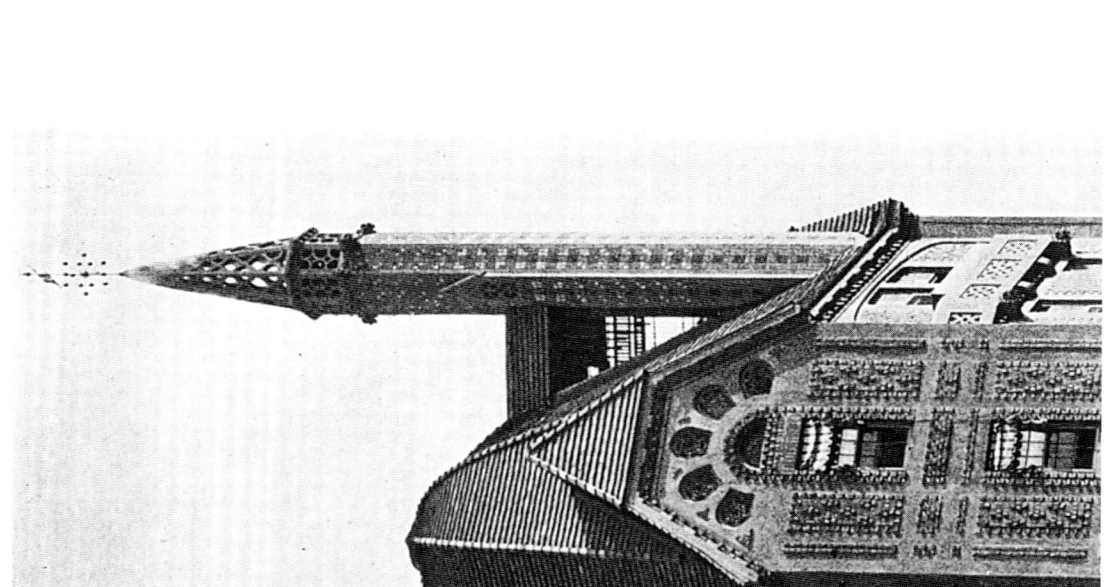

Fig. 212. PARRISH ART MUSEUM, SOUTHAMPTON, L. I.
Precast concrete tracery and concrete copings, both with exposed aggregate.
GROSVENOR ATTERBURY, ARCHITECT.

Fig. 210. ROOF AND FLECHE OF FOREST HILLS GARDEN TOWER

Precast, factory made concrete sections and decorative panels. Aggregates of dark blue and red tile were used for the Fleche. The inlays of concrete cubes were made of different colored concrete and the panels between the concrete "half-timber" work were laid up in brick. The openwork grilles inside the horseshoe arches were precast.

GROSVENOR ATTERBURY, ARCHITECT.

Fig. 213. CONCRETE TRACERY PARAPET OF A MEXICO CITY RESIDENCE
Prof. G. L. Zaccagna, Architect.

which it is poured it is no longer natural or necessary to have vertical or straight contours for wall-openings. Bricks, wood and steel beams are straight elements and therefore it has been natural to have windows and doors as well as other parts of a design rectangular, as the introduction of curves necessitated extra work. In concrete design curvilinear outlines must become the usual ones as they are more beautiful.

The wall of the medieval castle was support as well as protection against weather and attack; the small window-openings served for light, ventilation and shooting. The wall of the modern frame building consists of a supporting skeleton and protecting panels. In concrete buildings the panel can be further differentiated as an outside shell which bars wind and rain, and an insulating inner shell which, being porous, tends to keep out heat and cold and also can be nailed. These curtain-walls resemble in their constructive function the tympanon of the classical temple and the metopes of Doric friezes; we therefore would be following tradition in covering our concrete wall-panels with figures and ornament in tracery form. These panels are again divided into concrete strips which serve as bracing and railing, and into voids for lighting and ventilation. The final differentiation will consist in separating these last two functions: built-in window-panes such as those in the Veszprem theatre (where the ventilation openings are in the roof), and big windows with inserted ventilator panes point in this direction. The light-openings can have any de-

Fig. 214. CONCRETE TRACERY BALUSTRADE OF A RESIDENCE ON VIA DELL'USINA, GORIZIA, ITALY

Fig. 215. VILLA HOCKEGASSE, 71, VIENNA, AUSTRIA
Concrete tracery balustrade.
F. S. ONDERDONK, ARCHITECT.

sired, (also curvilinear) outline; one pane which can be easily opened will insure ventilation.

As Goethe pointed out, evolution follows a spiral curve, returning after a cycle to the starting point but on a higher level, and so concrete tracery harks back to the architecture of Egypt:

"The surface decoration of the Egyptian masonry walls is also held to have been derived from the pictures scratched on the early mud or plastered walls, which manifestly did not lend themselves to modelled or projecting ornament, though their flat and windowless surfaces were eminently suitable for incised relief and explanatory hieroglyphics."[45]

Concrete walls have two points in common with the ancient mud walls of Egypt: plasticity before setting and the desirability of avoiding projections—with concrete a matter of economical centring. On the other hand ferro-concrete is a hard, durable material which carries tension, and hence the pierced background of concrete tracery which gives a far more effective picture than the incised outlines of Egyptian wall decorations could.

Concrete tracery will help the architect to fulfill once more his su-

(45) Sir B. Fletcher, "History of Architecture", p. 21.

Fig. 216. RESIDENCE, BELMONT, MASS.
Concrete tracery balustrade reinforced with aluminum wire.
BATES & WIGGLESWORTH, ARCHITECTS.

148

Fig. 217. W. P. FULLER & CO. BUILDING, LOS ANGELES

preme duty—to create poetry and tell stories in stone—and to be heard. Concrete tracery with its black and white, eventually even with color, will convey the architect's message, and in a way to compete successfully for attention with the advertising signs which are at present the most conspicuous features of modern cities. L. N. Tolstoy, John Ruskin, and Upton Sinclair have demonstrated that the true artist always has a message, and concrete tracery will be an effective medium for proclaiming it.[46] As John Ruskin said:

"A wall surface is to an architect simply what a white canvas is to a painter, with this only difference, that the wall has already a sublimity in its height, substance and other characters already considered, on which it is more dangerous to break than to touch with shade the canvas surface."[47] "Better the rudest work that tells a story or records a fact than the richest without meaning. There should not be a single ornament put upon great civic buildings without some intellectual intention."[48]

These thoughts of Ruskin are so true and great that they need no further explanation—they just must be obeyed.

Many architects try to force three-dimensional beauty on city-house facades, which are essentially two-dimensional, by imposing orders on them; but as long as there is no space between colonnade and main wall the deep shadows are lacking and exactly

(46) "What is Art"; "The Seven Lamps of Architecture", "Mammonart".
(47) "The Seven Lamps of Architecture", 3-XII.
(48) "The Seven Lamps of Architecture", 6-VII.

Fig. 218. CONCRETE BALUSTRADE IN THE BATHING ESTABLISHMENT OF VOSLAU, AUSTRIA
Note the wavy outline of the balusters.

Courtesy *Concrete*
Fig. 219. CONCRETE TRACERY BALUSTRADE AND MONOLITHIC LAMP POSTS, FAIRMONT BRIDGE, W. VA.

the opposite effect of the one aimed at is achieved: there is no conquering of space, but a facade which has been forced into the defensive, and meanly capitulates before the building line. The zoning laws which have created the stepped-back facade for skyscrapers provide one solution; concrete tracery will give a second one: the facade is candidly declared a plane which is treated like a lace-curtain. The many voids of the tracery show a sufficient amount of the space behind the wall, dark by day, light at night, to give the three-dimensional impression of the *house* of which the facade is professedly only one plane. Instead of a portico degenerated into pilasters, the concrete colonnade is left in front and the intercolumnar spaces filled by grillework. The actual house, a dark or light space, remains behind this concrete lace-curtain, and the conquering of space is expressed, especially when a far-projecting cornice and bold balconies are used as finishing touches to the design. Concrete tracery would permit the architect to fully realize B. G. Goodhue's ideal:

"... not that the building should make place for sculpture after the manner of a pedestal or frame, but that the carven stone, whether image or inscription, should be of the substance of the building wrought at those points where relief could clearify the architectural form ... wholly within the structures and surfaces of the edifice."

Details of Execution.

With concrete tracery the fenestration can attain various degrees of ornamentation. The total window area can be adjusted to meet every demand. The concrete "picture-frame" enclosing glass but no concrete "picture" is the one extreme. In 1908 Professor Ernest Wilby, associated with Mr. Albert Kahn, designed the Ford Plant at Highland Park and

Fig. 220. FERRO-CONCRETE TRACERY BALUSTRADE, S. B. NEWBERRY RESIDENCE, CLEVELAND, O.

here for the first time the concrete frame itself was exhibited, the entire space from pier to pier being filled with glass (Fig. 236). Only below the sills did a low parapet of brick remain. The new treatment which had great practical and economic advantages spread rapidly over the country and has given the American industrial building its characteristic appearance.

In office buildings where the position of partition walls is not always decided on beforehand, concrete tracery would cover the whole facade in a simple geometric design, so that, wherever the partitions are placed, each room would have a completed window-group. In apartment-houses and residences the solid part of the tracery can expand until it occupies almost the entire wallspace and the total glass-area is equal to that of the customary windows (Fig. 232). Between these two extremes there is a wide range of intermediate stages

By Ewing Galloway, N. Y.
Fig. 221. OFFICE BUILDING, SAN JUAN, PORTO RICO
Concrete tracery balustrade.

By Ewing Galloway, N. Y.
Fig. 222. THE ATENO (ART CENTER), SAN JUAN, P. R.

From Ewing Galloway, N. Y.
Fig. 223. CONCRETE RESIDENCE, SAN JUAN, P. R.

Fig. 224. FERRO-CONCRETE SKELETON OF A HOUSE IN GENOA, ITALY

Fig. 225. MASONIC TEMPLE, WICHITA, KANS.
Eberson & Weaver, Architects.

(Fig. 237, 238). For small buildings concrete tracery of simple outlines, effective also from within as silhouette of dark on light, would be chosen (Fig. 243). Frank Lloyd Wright's residences (Fig. 165, 167, 239, 240) demonstrate by original window outlines and perforated textile-blocks ways of eliminating the customary tedious checkerboard design of rectangular windows. Simple types of concrete tracery are further employed in many buildings in California and Florida. The narrow panels with a grille of leaves on the pylons of the Miami Skottish Rite Temple designed by Kiehnel & Elliot (Fig. 245, 246) are examples of non-geometric concrete tracery.

In monumental public buildings the tracery of the large halls would depict historical scenes or a symbolic group of figures, arranged to be viewed from within as well as from without. The Midland County Court House (Fig. 247) contrasted with the Belgian House at the Paris Exposition of Decorative Arts of 1925 (Fig. 248, 249) shows that concrete tracery would be far more effective than frescos wedged between rectangular windows. Usually there would have to be a symmetry plane running thru the center of the tracery parallel to the facade which would be significant in case bas-relief as an addition to the silhouette-effect is desired. The sculptor could chisel before the concrete is quite hard or gluemolds could be inserted in the main centring. The inside wall surface, being so much nearer to view, would demand colored or mosaic covered concrete and possibly stained glass in the voids, as the figures would appear too large and be in shade.

Courtesy *Concrete*
Fig. 226. THE EAGLES LODGE, MILWAUKEE
Statues, frieze and ornament are concrete-stone.
RUSSELL BARR WILLIAMSON, ARCHITECT.

Classical art would furnish inspiration for the archaic rigor of the outlines of concrete tracery, which would be essential in making it clearly understood when viewed from the distance, for, according to John Ruskin, ".....the Greek workman cared for shadow only as a dark field wherefrom his light figure or design might be intelligibly detached: his attention was concentrated on the one aim at readableness, and clearness of accent; and all composition, all harmony, nay the very vitality and energy of separate groups were, when necessary, sacrificed to plain speaking."[49] Poster-art which has recently developed gives an example of the effect that can be achieved by simple contours and spaces of uniform tone relieved by darker ones.

The shadow intensity would be variable according to whether the window-panes are set nearer the outer or the inner wall surface. The voids could vary concerning the light they admit: stained glass, glass-stones as described below, or ordinary glass could be inserted, the latter varying in thickness and fluorescence. The solid part of the concrete tracery could be subjected to various kinds of surface treatment and by combining all the mentioned possibilities an almost unlimited amount of ornamen-

(49) "The Seven Lamps of Architecture", 3-XV.

Fig. 227. CZECHOSLOVAK BANK, VIENNA I., AUSTRIA

tal effects presents itself. Concrete tracery can be formed by placing thin strips of metal between the centering-shells. The glass panes might be placed on rebates in the concrete ribs as was done in the Veszprem Theatre, Hungary, or set like ordinary window-panes in lead-frames. Ventilation panes would be either pivoted, hinged or sliding.

The Luxfer-grille windows, a German innovation, are made by placing small precast concrete frames in parallel rows and pouring cement mortar in the joints. Areas larger than four square meters[50] need reinforcing bars in these narrow ribs. Various kinds of glass can be inserted and for insulation purposes double glazing is employed. The panes are attached with a special putty. Ventilation is taken

---

(50) The introduction of the metric system in the United States would help architects save time, money and energy. For information write to the **All America Standards Council**, 114 Sansome Str., San Francisco, Cal.

(51) "Wasmuths Monatshefte", Jan. 1928, p. XIV.

Fig. 228. HALL FOR FUNERAL SERVICE, CENTRAL CEMETERY, VIENNA, AUSTRIA
Aichinger & Schmidt, Architects.

care of by hinged units which are placed among the rigid ones.[51]

Glass-Stones.

"Translucent concrete" is the term adopted by a Frenchman to designate a concrete area into which hollow glass-stones or glass-bricks have been inserted. Glass-stones, as shown in the accompanying illustrations (Fig. 250-255), produce the decorative effect of a light area contrasting with the surrounding dull concrete. Tracery such as that of the window in the Bahai Temple (Fig. 209) lets the glass-pane serve as a dark background against which the tracery-bars stand out. But these German *Glasbausteine* and French *Briques Falconier*, themselves produce a remarkable effect thru their usually hexagonal shape. Their face sides have a series of stepped-back planes which cause the light to reflect and make the glass-area translucent instead of transparent. The final effect may be termed jewel-like. Falconier glass-stones are of dark

Fig. 229.
HALL FOR
FUNERAL SERVICE
CENTRAL CEMETERY
VIENNA
Interior view.
Aichinger & Schmidt,
Architects.

blue, yellow or of ordinary glass. The Dresden products are either of green or of plain glass.

An important use of non-structural glass-stones is for inside walls where they permit light to pass from an outside room into an otherwise dark passage. Their advantage over thin translucent glass panes is the enclosed air-space which facilitates insulation. The Electrotechnical Institute of the Vienna Technical University thus introduced light into one of its main passages. Glass-stones are

Fig. 230. FACTORY G. HUBBE-FARENHOTZ, MAGDEBURG, GERMANY

The network surface was created by means of a trowel in the stucco which covers this brick building. The triangular recesses were to have been left void as windows but other sources of lighting were used.

P. SCHAEFFER, ARCHITECT.

Fig. 231. BRIDGE OVER LENNE STREET, DRESDEN
(Erected for the International Hygiene Exposition, 1911).
The superstructure is timber-frame covered with stucco; the bridge proper and pylons are of ferro-concrete.

PROF. M. DULFER, ARCHITECT.

Fig. 232. DESIGN FOR AN APARTMENT HOUSE
Piero Portaluppi, Milan, Architect.

Fig. 233. DESIGN FOR AN APARTMENT HOUSE
Piero Portaluppi, Architect.

further inserted in exterior walls of factories, garages and other utilitarian buildings where light is needed but, due to fire-regulations, ventilation is procured by openings in other walls. Glass-bricks containing a wire mesh embedded in the glass are absolutely fireproof according to tests made at the Technical Universities of Dresden, Berlin and Munich.

Glass-stones are constructed with grooves; if at any time it is necessary to remove one unit, the producers claim it can be replaced without damaging the others. Mortar consisting of one part portland cement and three parts sand, and about ten per cent lime, binds the glass units. The joints between the hexagonal stones are 0.2", the longitudinal joints between the glass bricks only 0.11" and their side joints 0.4". After the walls are erected the joints are cleaned and pointed.

In Germany barrel vaults of glass-stones have been built by placing them like voussoirs between a frame of concrete arches and connecting horizontal beams (Fig. 254). Areas exceeding 10 sq. meters require steel rods in some of the cement mortar joints. In this vault colored glass-stones were inserted among the white ones, forming patterns.

The Vienna architect B. V. Nordenkampf used glass-stones in brilliant colors for columns and walls. Fig. 255 shows that they acted as centring and at the same time resulted in an attractive facing. The result was a glass mosaic of large units. In France unique results have been achieved by inserting glass-stones into concrete vaults of various types;

Fig. 234-235. EVANGELICAL GARNISON'S CHURCH, ULM, GERMANY
Professor Th. Fischer, Architect.

Fig. 236. ORIGINAL FORD FACTORY, HIGHLAND PARK, DETROIT, MICH.
Albert Kahn and Ernest Wilby, Architects.

Fig. 237. FACTORY G. HUBBE—
G. W. FARENHOLTZ, MAGDE-
BURG, GERMANY
Brick panels in concrete framework.
P. Schaeffer, Architect.

Fig. 238. ANATOMY PAVILION, UNIVERSITY OF MUNICH, GERMANY
Heilmann & Littmann, Architects.

symmetrically spaced they form patterns of light spots against a dark background. Thus the architect has a new, dignified, and effective mode of decoration at his command (Fig. 251-253, 346).

Pressed-glass bricks which act as bearing units are 25 cm.[52] long, 12.5 cm. high and 8 cm. thick. One type of hexagonal glass-stones made in Germany are 6 (14) cm. wide, 20 cm. high and 11 cm. thick.

A.-G. Perret's Concrete Tracery.

The essential suggestions of this chapter were advanced by the author in 1918 in a thesis submitted to the Vienna Technical University. Since then the brothers Perret, independently recognizing the same possibilities of ferro-concrete, have built several towers and churches featuring concrete tracery. Auguste Perret, born in Brussels 1874 and Gustave, born two years later, were sons of a builder and when, after studying at the Paris Ecole des Beaux-Arts they took over their father's business, they combined the architect's knowledge with the craftsman's practical experience as the Gothic masters had.

The Orientation Tower of Grenoble, France (Fig. 256-257) was built for the Water Power Exposition of 1925 at the instigation of the Touring Club which wished to mark this center of a "grand tourisme" region in a significant manner. The tower is 95.5 meters high and its base has a diameter of 7.95 meters. A few meters above the ground a marquise encircles the tower. The main shaft is terminated by a circular loggia formed by columns supporting a terrace on which the orientation chart-table is placed. This terrace is 60 meters above the ground and can be reached by elevators. The spire which surmounts this terrace has a central column supporting a spiral stairway and is crowned by the three roses of the Grenoble coat-of-arms. The panels which cover the shaft are pierced by triangular openings, a simple type of concrete tracery (Fig. 258).

The grilles in the tower of the St. Vaury church (Fig. 270) are built up of the same units as were employed

---

(52) 1 centimeter is one hundredth of a meter (about 2/5 of an inch).

Fig. 239. "LA MINIATURA," PASADENA, CAL.

First textile-block-slab house. (A combination of concrete walls and block-slab.) The concrete tracery above the windows acts as a very attractive wall-decoration on the interior (See the illustration in the "Architectural Record," Aug. 1928, p. 103).

FRANK LLOYD WRIGHT, ARCHITECT.

Fig. 240. FREEMAN HOUSE, LOS ANGELES, CAL.

Third textile-block-slab house. 16"x16 blocks.

"*As a plastic material—eventually becoming stone-like in character—there lives in concrete a great aesthetic property, as yet inadequately expressed.*" Frank Lloyd Wright, Architect. For the effect of the pierced blocks from within the house see the illustration on page 165.

Courtesy *The Architectural Record*
FREEMAN HOUSE, HOLLYWOOD, CAL.
Note the effect of the pierced blocks.
FRANK LLOYD WRIGHT, ARCHITECT.

Courtesy *The Architectural Forum*
Fig. 241-242. CORAL GABLES ELEMENTARY SCHOOL
Entrance Details.
KIEHNEL & ELLIOTT, ARCHITECTS.

in the all-concrete churches at Le Raincy and Montmagny, two suburbs of Paris.

Notre Dame du Raincy (Fig. 259-264), praised as the Sainte Chapelle of the Ferro-Concrete architecture, has a hall of 20 by 56 meters, with four rows of columns which are 11 meters high and 43 centimeters in diameter. The outside columns stand within the enclosing wall separated by a few centimeters of space so that the church "represents itself to us with divisions like the members of an organized body."[53] This detaching of the columns from the exterior wall which they ought to serve as framing seems disputable. The existing slope of the site was utilized to give the floor of the church an inclination similar to that of an auditorium, and to build under the raised apse a crypt containing the offices, sacristy, schoolroom and furnace-chamber. The ciborium which covers the altar consists of columns supporting angels cast of concrete in "geometric molds" and done over with gold. At the right and at the left of the entrance are polygonal chapels, the right one containing the baptismal font. Le Raincy church as well as the St. Therese church at Montmagny (Fig. 265-268) represent an important step in the development of concrete tracery as the entire wall, small areas excepted,

---
(53) Paul Jamot, "A. G. Perret et L'Architecture du Béton Armé".

Fig. 243. GRAMMAR SCHOOL, CORAL GABLES, FLORIDA
KIEHNEL & ELLIOTT, ARCHITECTS.

consists of a concrete grille composed of oblongs, squares, triangles, crosses, and circles which reoccur as perforations of the balustrades, vaults,[54] and tower-panels. These concrete blocks have coves around their edges and were cast in a few simple wooden molds.

There are no windows, as the term is usually understood, in these two churches, as the light-admitting glass fills the hundreds of little voids between the reinforced concrete network. Masonry as a pattern set off by a background of glazed voids is the epochal gift of ferro-concrete as these churches demonstrate. Whether the patterns in these particular examples are pleasing is a question of secondary importance. They must be considered as the primitive beginning of the most delicate and refined type of wall-treatment the world has witnessed. The price of the Le Raincy church explains the simplicity of the tracery: the architects had only approximately 600.000 francs at their disposal.[55] W. D. Foster gives the following description of the glass which fills the tracery:

"Naturally when there is so much glass surface the handling of the glass itself becomes of prime importance, and it has been done with the greatest care and subtlety, with the result that the light which floods the church at all times is astonishingly effective. In fact, it is so essentially a part of the whole effect that it is impossible to judge the atmosphere created

---
(54) The vault-perforations which are backed by heavy stuffs lessen the reverberation.

(55) Approximately $30,000 (?) in 1923.

Fig. 244. GRAMMAR SCHOOL, CORAL GABLES
Concrete tracery balustrades.
KIEHNEL & ELLIOTT, ARCHITECTS.

when one can merely see black and white reproductions. The colors of the glass nearest the entrance are of a general yellow tone, and as the windows approach the altar the colors become deeper, going through the oranges, the reds and violets and ending in blue for the space around the apse. This blue, with accents of red and violet interspersed, has the depth of a clear night sky. This, with the large pattern of a cross showing directly behind the altar, forms a solemn background for this most sacred part of a church. The windows along the sides, in their turn, have the individual units so arranged that a cross appears in the pattern of each one, while at the center of the cross is a decorative panel of glass representing not only biblical scenes but also scenes inspired by the Great War[56].... The panels are by the well-known painter, Maurice Denis."[57]

The tower of Notre Dame du Raincy is 43 meters high and is sup-

---

(56) The author regrets this adulteration of Christianity.
(57) "Architectural Record". Aug. 1924, p. 100.

Courtesy *The Western Architect*
Fig. 245. SCOTTISH RITE TEMPLE, MIAMI, FLA.
Improved concrete block and reinforced concrete framework, covered with portland cement stucco. Grills, ventilator and other ornaments of precast concrete. The columns are hollow and were cast in place. KIEHNEL & ELLIOTT, ARCHITECTS.

Courtesy *The Western Architect*
Fig. 246a. SCOTTISH RITE TEMPLE, MIAMI
Simple, square patterns were created by wooden inserts in the main shuttering; the eagles were modeled direct with cement, in position.
KIEHNEL & ELLIOTT, ARCHITECTS.

Courtesy *Architectural Record*
Fig. 246b. VILLA CARLOTTA, HOLLYWOOD, CAL.
Arthur E. Harvey, Architect.

Fig. 247. DETAIL OF MIDLAND COUNTY COURTHOUSE, MICH.
Magnesite stucco on celotex.
Bloodgood Tuttle, Architect.
Paul Honore, Painter.

Fig. 248-249. BELGIAN PAVILION, 1925 PARIS EXPOSITION OF DECORATIVE ARTS
Note the figural tracery frieze.
M. V. Horta, Architect.

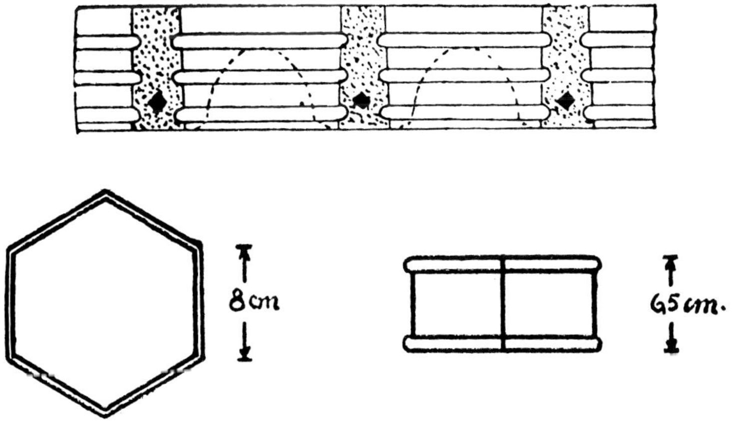

Fig. 252. LA BUTTE-AUX-CAILLES BATH, PARIS
Glass-blocks inserted in concrete.
L. BONNIER, ARCHITECT.

Fig. 250a. GLASS-BLOCKS BETWEEN FERRO-CONCRETE RIBS
Fig. 250b. "FALCONNIER" GLASS-BLOCKS

Fig. 251. GLASS-BLOCK FERRO-CONCRETE CEILING IN A PARIS STORE

"Once a precious substance limited in quantity and size, glass and its making have grown so that a perfect clarity of any thickness, quality or dimension is so cheap and desirable that our modern world is drifting toward structures of glass and steel."—FRANK LLOYD WRIGHT.

Fig. 253. POSTOFFICE, RUE BERGERE, PARIS
Glass-block ferro-concrete ceiling. F. LE COEUR, ARCHITECT.
"In the openings in my buildings, the glass plays the effect the jewel plays in the category of materials. The element of pattern is made more cheaply and beautifully effective when introduced into the glass of the windows than in the use of any other medium that Architecture has to offer."—FRANK LLOYD WRIGHT.

Fig. 254. MULTI-COLORED GLASS-BLOCK VAULT WITH FERRO-CONCRETE FRAMEWORK
"Shadows have been the brush-work of the architect when he modeled his architectural forms. Let him work, now, with light diffused, light refracted, light reflected—use light for its own sake—shadows aside. The prism has always delighted and fascinated man. The machine gives him his opportunity in glass. The machine can do any kind of glass—thick, thin, colored, textured to order—and cheap. A new experience is awaiting him."—FRANK LLOYD WRIGHT.

ported by columns of the same diameter as those carrying the vaults of the nave; these columns are assembled in groups of five and are more effective architecturally than a single pier of the same strength would be, and no more expensive. The customary wooden belfry which serves to protect the masonry of the upper tiers of the tower from the vibration caused by the ringing of the bells is omitted in this church, the bells being suspended from ferro-concrete beams which rest on the tower proper. The twenty columns which constitute the tower terminate at different heights, their number decreasing to twelve, eight, and four in successive tiers; the final central column supports a ferro-concrete cross. A French critic quoted by Paul Jamot compares the tower with its vertical lines and shafts of different heights to a monumental organ developed upwards instead of horizontally. In fact the lower part of this tower does contain the organ. In reviewing this church M. Maurice Brillant praises A. and G. Perret for having made of ferro-concrete "the most immaterial of mater-

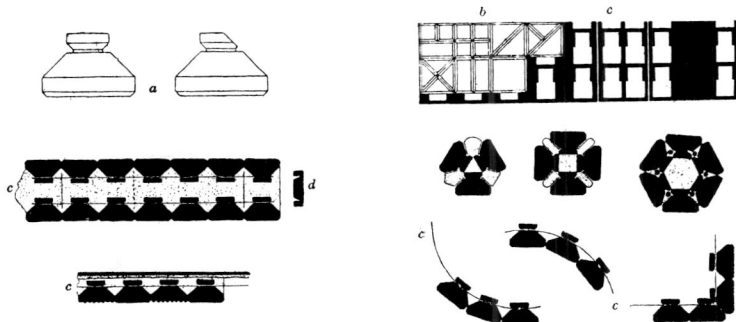

Fig. 255. SECTIONS OF COLUMNS AND WALLS MADE OF GLASS-BLOCKS INSERTED IN FERRO-CONCRETE
The glass-blocks which are shown black in this diagram are hung in metal frames "c" which are held in place by clips "d".
B. von Nordenkampf, Vienna, Architect.

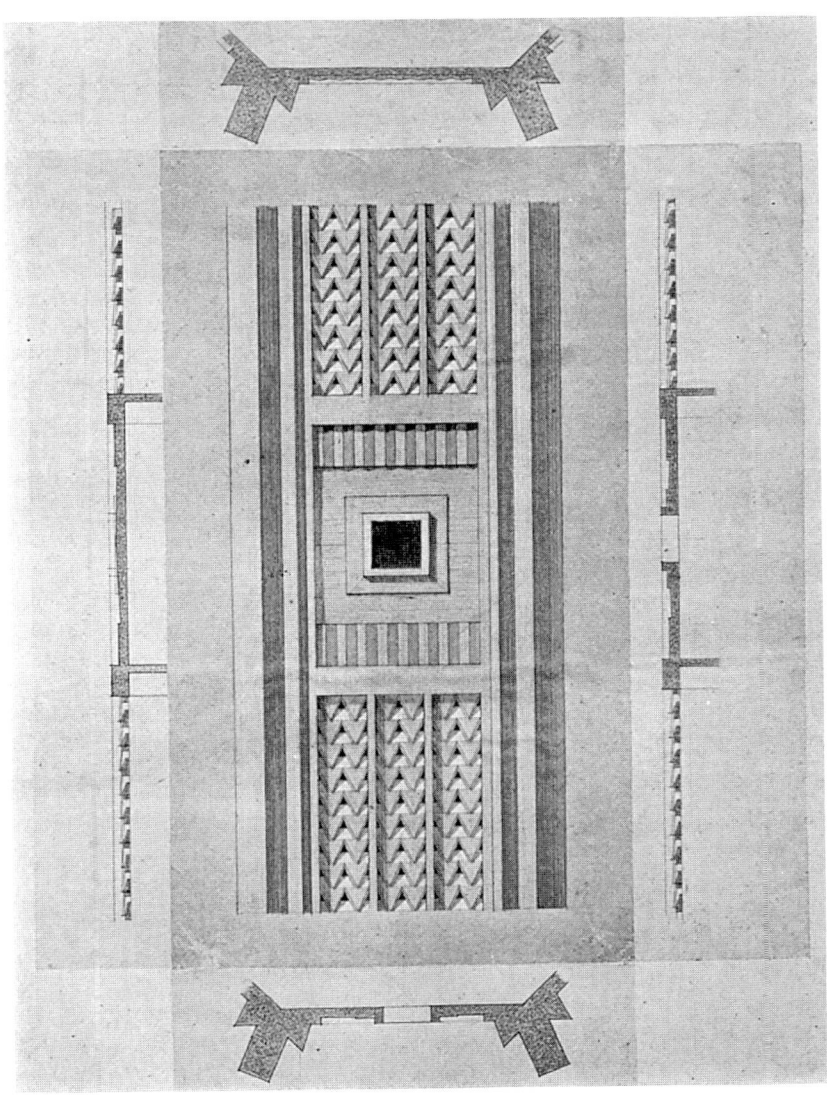

Fig. 256. ORIENTATION TOWER, GRENOBLE, FRANCE
Sections thru tower and detail of concrete tracery panel.
A. & G. Perret, Architects.

 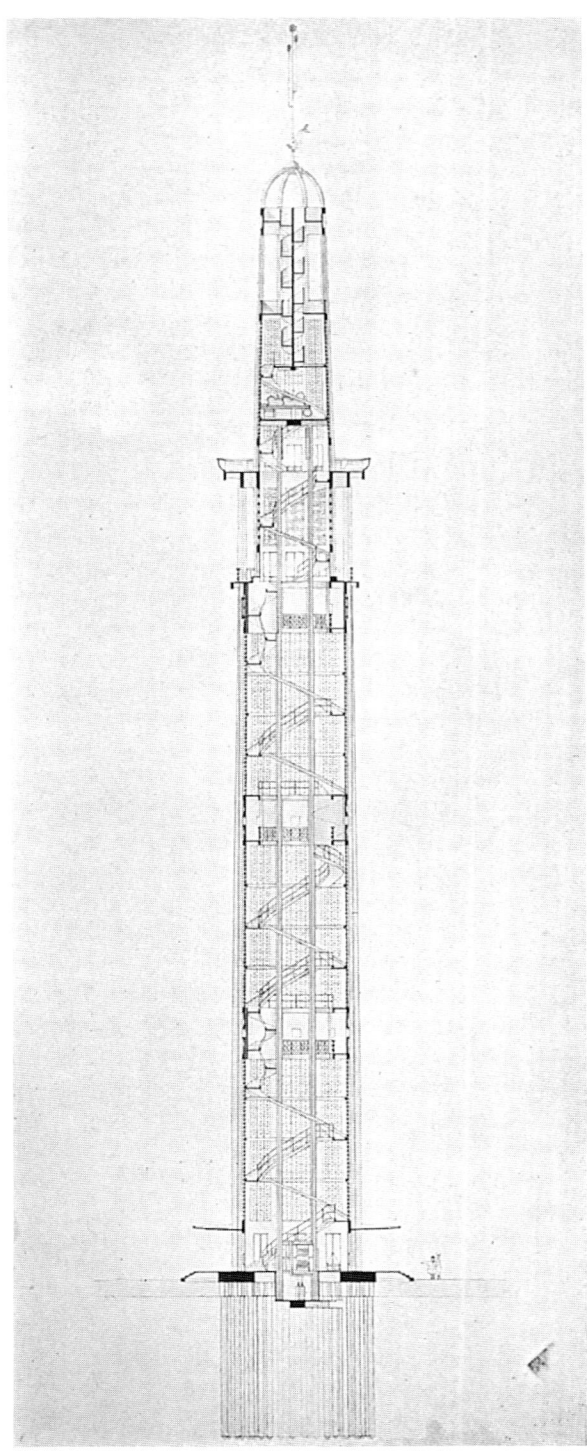

Fig. 257-258. ORIENTATION TOWER, GRENOBLE, FRANCE
A. & G. PERRET, ARCHITECTS.

ials". Paul Jamot terms Notre Dame du Raincy an alliance of the solidity and rational geometry of Classicism with Gothic daring and spirituality, which union is based on the verticals and rectangles which he conceives as characteristic of ferro-concrete.

St. Therese at Montmagny, completed in 1926 is not surrounded by buildings, and hence is flooded with light. The light shines thru the many openings in the concrete tracery of the farther wall, piercing the colored panes of the nearer grille and thus showing their brilliant hues on the *exterior*. The church has the effect of a colored lampshade and the architect who beholds this novel spectacle realizes that St. Therese foreshadows a revolution in architecture. He visions the plain circles and squares changing into intricate curves and outlines of figures, and the yellow and blue panes of this church being replaced by the rainbow of Gothic stained glass: the monumental building of the fully developed Ferro-Concrete Style, whether museum, library, or church, stands before him as a picture of transparent jewels set in curvilinear concrete frames.

The brothers A. and G. Perret prefer to leave the concrete visible because "well worked concrete and especially polished concrete can acquire an aspect comparable to that of granite, the most beautiful, hard, and durable of all materials".[58] P. Jamot calls economy the supreme rule in Architecture and states that A. Perret's esthetic is contained in the motto: "Do the best possible with the least possible amount of material and labor." Jamot adds that concrete, which per-

(58) Paul Jamot.

Fig. 259. NOTRE DAME DU RAINCY, PARIS
A. & G. Perret, Architects.
(1922-23)

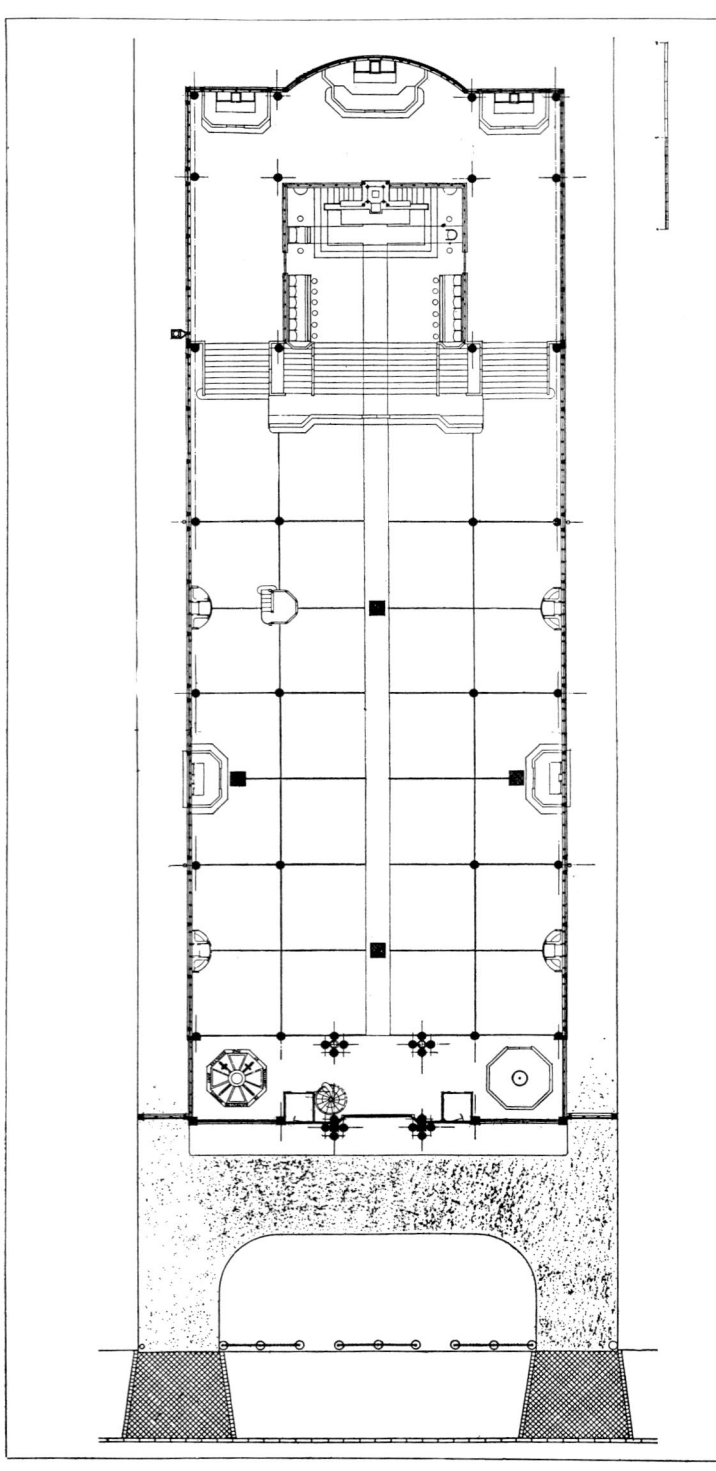

Fig. 260. NOTRE DAME, LE RAINCY, PARIS
Plan.
(The 4 black squares represent floor-grilles).

mits the building of indestructible vaults which are comparable in their thinness to an egg-shell, and which replaces the wall by a few columns, lends itself better than any other material to this classic conception of art.

The regrets of M. Jamot that the Jury did not choose A. and G. Perret's design for the church of St. Jeanne D'Arc (Fig. 269) will be shared by all pioneers of the Ferro-Concrete Style. In this design A. Perret was guided as in his previous ones by Fenelon's principle: "No part of a building should be devoted to ornament only; rather, aiming always at beautiful proportions, all the structural elements of a building should be turned into ornaments." The dominant feature of the design is the 200-meter-high tower to be built of four piers, each of which in turn to consist of four columns. Up to a height of 110 meters glazed grilles would fill the panels. The grille-walls of the church were to have two shells enclosing an airspace 60 to 70 centimeters wide, which would serve as insulation and maintain all parts of the wall accessible.

Paul Jamot sees in the central tower a spiritual symbol—arms extended to God; he believes that this edifice would fulfill the dreams of the Gothic builders who, due to the weakness of stone, were not able to erect as many nor as high towers on their cathedrals as they had planned. He cites Beauvais cathedral as the outstanding example of brittle stone failing the Gothic designer in his striving for the very heights which A. and G. Perret could easily attain with reinforced concrete. In complaining of the prejudices still existing in France against reinforced concrete and its principal

Fig. 261. NOTRE DAME, LE RAINCY, PARIS
Interior.

Fig. 262. NOTRE DAME, LE RAINCY, PARIS
Apse.

champion, A. Perret, he mentions that for a long time neither in Germany nor in France was there anything to indicate the birth of a truly architectural esthetic based on reinforced concrete. Jaul Jamot goes so far as to claim that A. Perret created Ferro-Concrete Architecture.

It should be realized that former generations were not able to replace the wall completely by a tapestry of tracery because stone-tracery lacked the reinforcing bars which, like concrete, can twist and turn into every imaginable curve and shape. We must follow the eternal principles of true art set forth by the masters of the past, at the same time fully utilizing the marvelous possibilities afforded our generation by ferro-concrete. Concrete tracery promises to be the glory of the Ferro-Concrete Style. May it not be that it is one of the essentials for which modern architecture has been waiting?

Fig. 263. NOTRE DAME, LE RAINCY, PARIS
Rear Facade.
A. & G. PERRET, ARCHITECTS.

Fig. 265. ST. THERESE, MONTMAGNY, (near Paris)
Side Facade.
A. & G. PERRET, ARCHITECTS.

Fig. 266. ST. THERESE, MONTMAGNY, FRANCE
Interior.
The stair leads to the organ.

Fig. 267. ST. THERESE, MONTMAGNY

Fig. 268. ST. THERESE, MONTMAGNY
Interior seen from the altar.

Fig. 269. DESIGN FOR THE JEANNE-D'ARC MEMORIAL CHURCH
The Apse in foreground.
A. & G. Perret, Architects.
(1926)

Fig. 270. FERRO-CONCRETE TOWER BUILT ON THE ST. VAURY CHURCH
(9 Xth Century).
A. & G. Perret, Architects.

CHAPTER IV
THE PARABOLIC ARCH

Fig. 271. PLANETARIUM
(Rheinhalle), DUSSELDORF
Lobby.
(See Fig. 29)
ALLGEMEINE HOCHBAU-
GESELLSCHAFT, A. G. BUILDERS.
PROFESSOR WILHELM KREIS,
ARCHITECT.

The manner in which two columns or walls are bridged over is the chief characteristic of a style. The Egyptians and Greeks laid a beam from post to post; the Romans adopted this post and lintel construction, but also connected two columns by an arch and two walls by a vault as the Assyrians and Etruscans had done before them. Later Gothic replaced the round arch of the Romanesque style by the pointed arch. Our wooden frame house, as well as the modern steel-frame building, again returns to the principle of post and lintel construction. The question therefore arises, as to how ferro-concrete will solve this problem. Buildings consist-

Fig. 272. HAT FACTORY,
LUCKENWALDE, GERMANY
Coloring Plant.
E. MENDELSOHN, ARCHITECT.
(1921-1923)

186

Fig. 273. HAT FACTORY FR. STEINBERG, HERRMAN & CO., SPINNING MILL, LUCKENWALDE, GERMANY
ERICH MENDELSOHN, ARCHITECT.

ing of one big hall will be treated in this chapter and those with several stories in the following one.

All the traditional methods have one characteristic in common: they piece stones, bricks, wood, or steel together by mortar, rivets, nails or bolts. Ferro-concrete is unique in creating buildings which form one solid mass; it has no "seams", once the sand, gravel and steel have been cemented together. In reinforced concrete columns and beam, walls and vault, vertical and horizontal become one. For this reason the rigid bent which in Europe is used so frequently for large halls is typical of concrete. This concrete portal type has structural and economic advantages due to the continuity of bending. Some designers place such a frame on hinges while others connect it rigidly with the foundation. The hinged type has columns tapering downward, i. e. growing narrower towards the floor because stiffness must be increased at the top and there the column is widest (Fig. 271-274, 300). W. W. Clifford, an American engineer, recommends the rigid bent in garages and similar buildings for spans of up to 75 feet where first class construction without columns is required as it is cheaper than a fireproofed truss, and in the shorter spans cheaper than the exposed truss.[59]

As in former transition periods the old forms are resurrected in the new

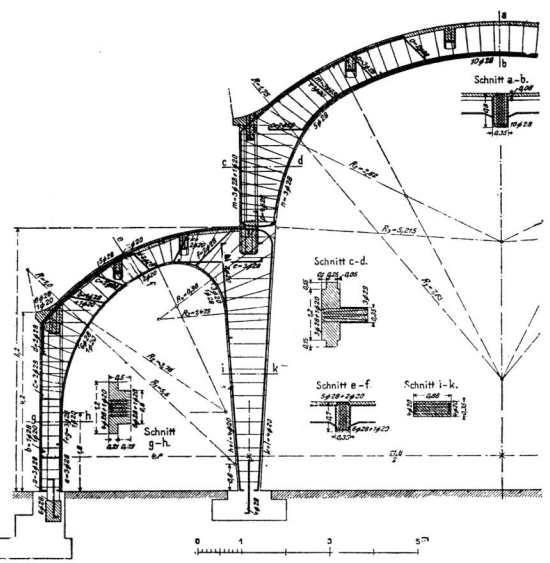

Fig. 274. NEW CATTLE HALL, OSNABRUCK, GERMANY
Transverse section.

---

(59) "The Architectural Forum", Sept. 1922, p. 148, 149.

187

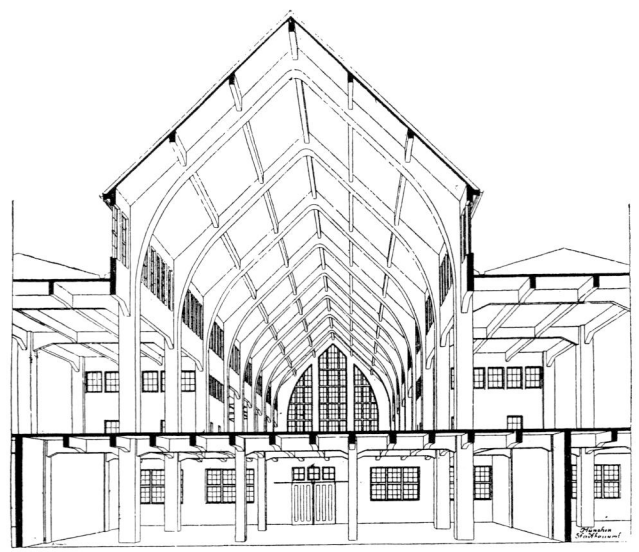

Fig. 275. GREAT MARKET HALL, MUNICH, GERMANY

material. The now dominating type of frame, two posts connected by a beam, is reminiscent of the old materials, just as the V-frame (Fig. 275, 44, 101) is due to the traditional type of roof. Gravity demanded vertical piles of stones or bricks to avoid centering; wood and steel encouraged angular outlines, but "liquid stone" requires no straight lines. Its form can express the structural facts: sides and top, walls and roof are one. The curved h a u n c h (Fig. 273-277) is much more pleasing than the angular one (Fig. 271, 272, 278) which is a product of the "wood-centring style." O. C. Hering considers the flattened arch with brackets at the point of juncture of beam and column the type that replaces the straight horizontal beam and the semicircular masonry arch.[60] This is only true for buildings with several stories. In the

(60) p. 102.

Fig. 276. COVERED LUMBER-COURT, KASSEL, GERMANY
Kurt von Brocke, Architect.

Fig. 277. CARPENTER-SHOP, HENSCHEL & SON LOCOMOTIVE WORKS, KASSEL, GERMANY

Fig. 278. COALBUNKER, KASSEL, GERMANY
Tooled concrete surface.
Kurt von Brocke, Architect.

hall-type building an arch or vault that springs from the ground, serving in its lower parts as wall, and in its top part as ceiling is characteristic of the Ferro-Concrete Style (Fig. 279, 346, 347, 380-384). The parabolic arch which near the base is almost vertical and then curves out gradually more and more till at the top it approaches a semi-circle is the ideal solution (Fig. 280).

The parabolic arch is both logical and beautiful. Hundreds of concrete bridges show its dynamic grace (Fig. 281-289). If spanning a void is the main problem of architecture, bridges may be considered to show us the ideal solution. For where could a form undergo a severer test as regards its esthetic value than when placed for a utilitarian purpose in a natural set-

Fig. 279a-b. PRUSSIAN MINING ACADEMY, CLAUSTHAL-ZELLERFELD, GERMANY
Auditorium.

Designed for ferro-concrete but executed in steel and stucco on metal lath. The hollow columns have a favorable acoustic effect. The hall is lighted indirectly from above.

REGIERUNGSBAUMEISTER ROTHER, ARCHITECT.

Fig. 280. MARKETHALL, BRESLAU, GERMANY
Dr. Eng. Kuster, Architect.

Fig. 281. QUAINT BRIDGE OVER THE REDNITZ
BY FURTH, BAVARIA
BROS. RANK, ARCHITECTS.

ting of rhythm and serene beauty? The parabolic bridge has become the standard type, not only because it is the most economical one, but also because the engineer senses its beauty. If it be acknowledged that sincerity is an essential element of the beautiful, it is significant that bridge builders prefer the parabola. Brick and stone construction would be too cumbersome, steel too skeleton-like to produce such elegant, thrilling arches as those shown in Fig. 281-289. Only ferro-concrete, combining the advantages of steel and stone, can do this.

Structural Advantages of the Parabola

Parabolic arches are economical: with a given load they require the smallest amount of material. The thrust curve of an arch carrying an evenly distributed load,—and that is the most frequent type of load in a building,—is a parabola. A parabolic arch can have a smaller section and less reinforcing than any other arch to support a given load. To quote from J. C. Austin:

(61) "Proceedings of the Am. Conc. Inst.", 1927.
(62) "Handbuch f. Eisenbetonbau", III-Vol. 2, H. 3.

"The use of the principles of the arch, particularly in reinforced concrete, should be exceptionally well adapted to our structures. It is a fact, that, generally speaking, no more than 37½ per cent of the concrete in a straight beam or girder is figured to resist compression, whereas close to 100 per cent of the concrete in an arch is usually effective in resisting compressive stresses. The arch is one of our most economical structural units."[61]

Professor C. Korner recommended that vaulted halls be more frequently designed to follow a parabolic curve because of all simple curves the parabola alone designates the vault of greatest stability, as investigations have proved.[62] The designer of the Breslau Market hall (Fig. 280) chose the parabolic arch because it approximated the line of thrust best. If it be admitted that architecture is to a large extent the symbolic expression of the conflict of forces acting in a structure, then the parabola is unquestionably to play an important role. The parabola can be called the resultant between gravity and a constant horizontal force. Gravity creates a parabola when drawing water earthwards, be it squirted from a horizontal spout or thundering over the Niagara Falls; gravity creates a catenary—a curve very similar to a parabola—when acting on a flexible material of uniform weight such as a chain or a ribbon. Plumet designed the outline of the Lalique fountain for the Paris Exposition of Decorative Arts (Fig. 290) as a slender parabola,—hence in harmony with the curves of the water-jets that spouted forth in all directions from various levels. The engineer creating a bridge and the architect designing

Fig. 284. CONCRETE BRIDGE WITH CAST IRON CORE, GMUNDEN, UPPER AUSTRIA
Span 70 meters.
Dr. F. von Emperger, Eng.

Fig. 285. FERRO-CONCRETE BRIDGE, SCHARNSTEIN, UPPER AUSTRIA
Dr. F. v. Emperger, Eng.

Fig. 282. OLD STEEL BRIDGE EMBEDDED IN CONCRETE, VELANDA, SWEDEN

Fig. 283. BRIDGE AND POWERHOUSE OF THE MITTLERE ISAR A. G., FINSING, BAVARIA

193

Fig. 287. CAPPELEN MEMORIAL BRIDGE OVER THE MISSISSIPPI RIVER, MINNEAPOLIS

Fig. 286. WASHINGTON MEMORIAL BRIDGE, WILMINGTON, DEL.
Note the "springy" effect of the central parabolic arch in contrast to the adjacent ones.

Fig. 288. FERRO-CONCRETE BRIDGE IN SWEDEN

a lofty hall simply "listen in" to nature when they choose a parabolic arch.

The Esthetic and Symbolic Value of the Parabola

Architects are inclined to disagree when the question of beauty arises. An object that is highly praised by some, is condemned as ugly by others. It will be therefore helpful to investigate the esthetic qualities of the parabola.

Greek art, generally accepted as the highest standard, created the sections of its moldings as approximate parabolas, hyperbolas and ellipses; the echinus-section of the Poseidon Temple in Paestum, a half-parabola, is one example. The parabola is a curve of changing curvature,—according to John Ruskin a characteristic of all beautiful curves.

The parabola is a curve that never ends,—one end of an infinitely long ellipse; comets follow parabolas thru limitless space and the parabolic arch gives the Ferro-Concrete Style something of the infinite swirl and swing of the Universe. To be "in tune with the infinite" has always been man's longing; but our age especially admires the Great, the Limitless. Lindbergh's flight enraptured mankind because he conquered space—pushed forward the boundaries in man's struggle with the infinite. Technical achievements and records established at the Olympic games satisfy man's longing to overcome his limits. This was recognized by Popper Lynkeus, a noted Austrian engineer and philosopher, when he wrote:

"Man's thirst for esthetic delights reveals itself as an expression of his reaction

Fig. 289. FORD BRIDGE OVER THE MISSISSIPPI RIVER

against temporariness, as desire for happy eternity." "Fundamentally engineering is concerned with the creation of contacts with surrounding nature, with the enlargement of the realm of power of our organs and nerves... a new conception of space and form-creation develops, a new idea of architecture, a new idea of beauty."[63]

The rhythm of rail and wire has already inspired poets and painters, and we might trace something of this modern estheticism in the hobo's passion for traversing the continent as "blind" passenger of our railroads. The architectural expression of this new rhythm, of this urge toward the infinite is, I believe, the parabola—the curve that pays us a visit en route from the infinite to the infinite. As Plotin's philosophy claims—"the apprehension of the infinite in finite figures can produce the impression of serenity... We have a longing born of an infinite and directed to an infinite desire of the soul which therefore never can fathom itself." We can call the parabola the emblem of man's life: one end here in the finite —the other hidden in the infinite-eternal—for are we not like comets of hidden origin and unknown destiny? Professor Durant calls our life "our parabola thru the world" and A. von Harnack writes—"But where and how the curve of the world and the curve of our own life begins— that curve of which it shows us a segment—and whereto this curve leads, science tells us nought."

Pythagoras considered the realization of simple numerical laws as the source and the essence of the soul's satisfied pleasure in contemplating the beautiful and the true: the parabola's equation is simple.[64] Plato recognized beauty as such in those simple geometric figures which affect us not by material charm but as representations of spiritual concepts, and Plotin states—"always that is pleasing which expresses ideas as purely and perfectly as possible ... The beautiful object becomes beautiful thru the dominance of the spirit, the dominance of the ideal over sensual matter, thru participation with reason which emanates from the Divine...."

The idea shining thru matter is the beautiful according to Hegel. An-

(63) Lux, "Ingenieur Aesthetik".
(64) $Y^2 = 2PX$.

Fig. 290. LALIQUE FOUNTAIN, PARIS EXHIBITION OF DECORATIVE ARTS, 1925
Architect, Ch. Plumet.

other German philosopher, Th. Fechner, applying these thoughts to the conical sections[65], called the parabola the symbol of love for the infinite and divine. He compares the focus to a soul, and the rays, emanating from it to the periphery, to the endeavors of this focus-soul, adding:

"The parabola is a serene symbol of love for an ideal, to the nonsensual, to all great and beautiful which—only attainable in infinity—, entices the soul: all rays sent out by the parabola-focus run in parallel direction to the other focus in the infinite; all desires and thoughts are only directed thereto. On the other hand no ray which did not emanate from the infinite can fall into the soul...... As egotism relates everything to self (symbolized by the circle) the parabola does everything without reference to itself as the ray must first travel to the infinite ere it can return to the focus; for this reason the parabola is also the symbol of virtue which, only by affecting the universe, wants to act for and on itself—and the symbol of virtue coincides with the symbol of love for the infinite."

(65) A plane cuts a cone either by a circle, an ellipse, a parabola or a hyperbola according to the angle which the axis of the cone forms with the cutting plane.

In adopting the parabola the Ferro-Concrete Style would resuscitate the Gothic tradition which Woringer terms—"mathematics come to life," adding:

Courtesy *Gleason-Tiebout Glass Co.*
Fig. 291. CELESTIALITE EXHIBIT AT THE ARCHITECTURAL AND ALLIED ARTS EXPOSITION, N. Y.
Howard Greenley, Architect.

Fig. 292. HANGAR FOR DIRIGIBLES, ORLY, FRANCE
Freyssinet, Eng.

".....Gothic architecture represents a complete dematerialisation of stone and is full of intellectual expression unlimited by stone and the senses .. for as soon as we receive a line in our consciousness we inwardly follow its generation. Viewing a Gothic cathedral we lose the feeling of our earthen shackles, we submerge in an infinity-movement which extinguishes all consciousness of the finite....The accentuation of the vertical produces here the infinity symbol."

Hogarth perceived that certain lines to be aesthetically fruitful must appear as nascent. For this reason the parabolic arch is most effective when it seems to spring from the ground as in Professor Bohm's church (Fig. 314-319); when it appears to shoot forth from the earth its infinite character is best felt as in parabolic bridges.

Examples.

Several large halls which utilize the parabolic arch have been erected in Europe. The airship hangar at Orly (Fig. 292) consists of a series of parabolic arches connected by horizontal bars forming oblong windows between. The noted German architect Professor Kreis designed a pavilion for the Munich "Farbenschau" (Fig. 293) featuring an arcade of parabolic arches which demonstrate the charm of unity possessed by that curve.

In the interior of the Engelbrekt church, Sweden, Professor Wahlman has made use of the parabola in a group of arches supporting the roof over the four arms at the crossing (Fig. 295-296). He explained that he desired to produce an effect of springiness such as it seemed impossible to obtain otherwise.[66] These arches are not of concrete, tho other parts of the church are. The parabolic arches at the crossing of St. John the Divine, New York (Fig. 298) are of huge granite blocks and designed to support a tower; it is significant that they

---

(66) Compare the illustrations of this church with Fig. 201-207.

Fig. 294. VILLA NUSSWALDGASSE, VIENNA XIX
Prof. Josef Hoffmann, Architect.

Fig. 293. EXPOSITION BUILDING AT MUNICH "FARBENSCHAU", 1923
Professor W. Kreis, Architect.

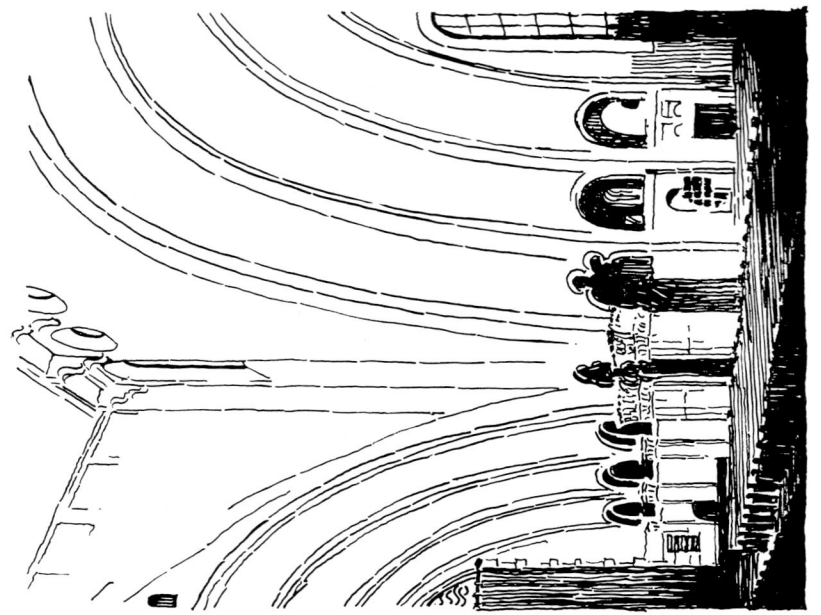

Courtesy *The Western Architect*
Fig. 296. ENGELBREKT CHURCH, STOCKHOLM
SKETCH BY ERNEST O. BROSTROM.

Courtesy *Charles Scribner's Sons*
Fig. 295. ENGELBREKT CHURCH, STOCKHOLM, SWEDEN
PROF. L. I. WAHLMAN, ARCHITECT.

200

Courtesy *The Western Architect*
Fig. 297. DESIGN FOR A PROTESTANT CHURCH
GERHARD PUSCH, ARCHITECT.

Courtesy *The American Architect.*
Fig. 298. CATHEDRAL OF ST. JOHN THE DIVINE, NEW YORK
Note the parabolic arches at the crossing.
CRAM & FERGUSON, ARCHITECTS.

Fig. 299c PALLOTINER CHURCH, LIMBURG L., GERMANY
NAVE AND CHOIR
J. H. Pinand, Architect.

SIDE CHAPEL

CLOISTER

PULPIT

Fig. 299 d-f PALLOTINER CHURCH, LIMBURG A. D. LAHN
J. H. PINAND, Architect.

Fig. 299a. CONGRESS HALL, GOTHENBURG, SWEDEN
Timber-construction on concrete foundations.
J. Lindberg, Eng.

Courtesy *The Architectural Record*
Fig. 299b. CHURCH IN LIMBURG, GERMANY
J. H. Pinand, Architect.

have been incorporated in this otherwise medieval design. The Congress Hall at Gothenburg, Sweden, (Fig. 299) is covered by a series of arches which support stepped-back vertical windows—an arrangement used with a circular plan in the gigantic Centenary Hall in Breslau, Germany (Fig. 380-384). The Exposition Hall in Magdeburg (Fig. 300, 301) and the Gymnasium of the Muhlheim High school (Fig. 302) show how much more graceful the arched support is than the angular frames so often employed for roof-construction (Fig. 271, 336).

The concrete tower placed on the old church of St. Vaury (Fig. 270), consists of four converging semiparabolas with concrete tracery between, which shows that the Perret brothers have sensed the two main characteristics of the Ferro-Concrete Style.

For small dwellings also, the parabolic vault has demonstrated its usefulness: the walls of the English Cat-Ar-System houses curve gradually into the roof so that the entire shell of the house forms a parabolic vault (Fig. 337). This vault consists of a series of arches made of concrete blocks which are held in position by the insertion of expanded metal reinforcing that provides sufficient adhesion. These arches carry only their own weight, the joists having their bearing on the end walls. The foundations are constructed in such a manner that spreading of the arches is prevented. No rainwater pipes or gutters are used, the water being conducted into channels in the ground, from which it runs into the sewer.

Details of Execution.

Everything new creates prejudice; yet the parabolic arch is not new. In the rare cases when the Egyptians used vaults, they preferred the parabolic arch to a semicircular one, the apex of which would have been more likely to yield under pressure, especially as mud bricks would be more

Fig. 300-301. EXPOSITION HALL "STADT UND LAND", MAGDEBURG, GERMANY
Bruno Taut, Architect.

Fig. 302. GYMNASIUM IN THE HIGH SCHOOL OF MUHLHEIM, GERMANY

Fig. 303. POSTOFFICE, UTRECHT, HOLLAND
J. Crowel, Jr., Architect.

Fig. 304-305. CHURCH IN NIJMEGEN, HOLLAND
Ir. H. Thunnissen, Architect.

easily crushed than kiln bricks.[67] The domes of the Babylonians and the gigantic vaults and cupolas which covered the Sassanian palaces approached the parabolic shape rather closely. Again the parabolic arch was used by the Saracens and later determined the shape of some of the Baroque domes. Yet it is so unusual in western architecture as to create various doubts in the minds of architects.

In Holland quite a number of buildings with vaults, and arched doors and windows of approximately parabolic outline have been built recently (Fig. 303-313). For insulation purposes concrete walls usually have an airspace, and this could be utilized in arranging parabolic sliding doors and windows which would be shoved into the slit between the concrete shells. Larger doors can be arranged as revolving doors. Parabolic windows could also receive an upright center-post around which the two wings would turn on hinges, while the upper segment would turn around a horizontal transom bar.[68]

The circular segment at the crown of the parabola admits more light and air than the point of the Gothic arch. When still more light near the ceiling is needed, the spandrels above the arch could be pierced and filled with tracery, or receive vertical posts of the bridge-type (Fig. 284-289), which transmit the load of the beam to the arch.

An arcade of parabolic arches creates piers which grow narrower toward the bottom due to the ever in-

---

(67) Flinder Petrie, quoted in Sir B. Fletcher, History of Architecture. p. 39.
(68) Two windows of this type were built in the house shown in Fig. 215.

Fig. 306-307. CHURCH IN ROTTERDAM
The arches of this and the Nymegen church are approximately parabolic.
Ir. H. Thunnissen and P. Buskens, Architects.

Fig. 308. APARTMENT HOUSE, REELOF HARTPLEIN, HOLLAND
Note the parabolic doors at the left and right.
G. J. Rutgers, Architect.

Fig. 309. TWIN VILLA, HILVERSUM, HOLLAND
Note the arches in the garden-gates and the main doors.
J. Van Laren, Architect.

creasing width of the parabola,—and this conflicts with the demand that a supporting pier taper towards the top. By setting the arches and their spandrels a few inches back of the front plane of the piers, the latter can stand forth, either with vertical contours, or growing wider towards the ground. Another solution is shown in Fig. 293.

We are so accustomed to the vertical posts supporting a beam or a semi-circular arch, that the slight slant in the lower portions of the parabolic

Fig. 310. BOY SCOUTS' CLUB-
HOUSE, LAREN, HOLLAND
J. Van Laren, Architect.

arch seems an obstruction. This may be just prejudice. The Mayan architecture utilized corbelled arches which created apertures that grew narrower towards the top (Fig. 39, 44, 141-144).

The hinged rigid frame as discussed in the beginning of this chapter requires likewise a widening of the column towards the top. The new Pleyel Concert hall in Paris has tapering walls due to this construction.

The Cat-Ar houses in England (Fig. 337) have built-in closets in the lower portions of the wall, thus utilizing the increasing width of the parabolic roof-wall. Some European theatres have seats quite close to the base of the parabolic arches so that the aisle is pushed away from the wall and has sufficient head-room, without wasting space. These examples show that the designers must exercise their ingenuity to turn what appears to be a disadvantage of the parabolic arch into a virtue. Further the proper relation of span to height must be carefully considered when design-

Fig. 311. COUNTRY HOUSE
"CORVUS", HOLLAND
J. Van Laren, Architect.

Fig. 312. SHOPS, HILVERSUM, HOLLAND
J. Van Laren, Architect.

ing a parabolic vault or arch. If construction, economy, and esthetics direct us to use the parabola, the slight slant of the lower portions which replace the vertical columns and walls should not seem too important, as we gain a wonderful unity and elegance for the outline of the opening (Fig. 280).

Only the applications shown in Fig. 280 and 302 were known to the author in 1918 when writing these suggestions concerning the parabola.[69] The parabolic arches and vaults which have been built since then have strengthened his conviction that the parabola is the characteristic curve for the Ferro-Concrete Style.

Prof. D. Bohm's Churches.

Professor Dominikus Bohm, head of the Department for Religious Art of the Cologne Crafts School, has created an outstanding example of a parabolic vault in the Catholic Church at Bischofsheim (Fig. 314-

---

(69) "Die Moglichkeiten einer Eisenbeton Architektur"—Dr. Dissertation.

Fig. 313. SHOPS, HILVERSUM, HOLLAND
Show-Windows.
J. Van Laren, Architect.

Fig. 314. CATHOLIC CHURCH, BISCHOFSHEIM, GERMANY
Nave.
Prof. D. Bohm, Architect.

Fig. 315. CATHOLIC CHURCH, BISCHOFSHEIM, GERMANY
Tower and Front-Facade.
Prof. D. Bohm, Architect.

319) which reminds of Fechner's statement concerning the "curve of idealism". This church as well as the unique one at Neu-Ulm (Fig. 320-331) mark Professor Bohm as one of the leading architects of the Ferro-Concrete Style.

The Bischofsheim church is well evaluated by the words of a German architect: "The piers shoot upwards, volcano-like;—colossal stress,—homogeneous mass." This church may appear too dark, but inserted glass-stones would create a parabolic vault that is flooded with light (Fig. 254). Had the door of the Bischofsheim church been parabolic like the tower-windows, it would have harmonized with the interior arches.

The Neu-Ulm church is especially remarkable as it it built without molds by "permanent centering" as described in the first chapter. The

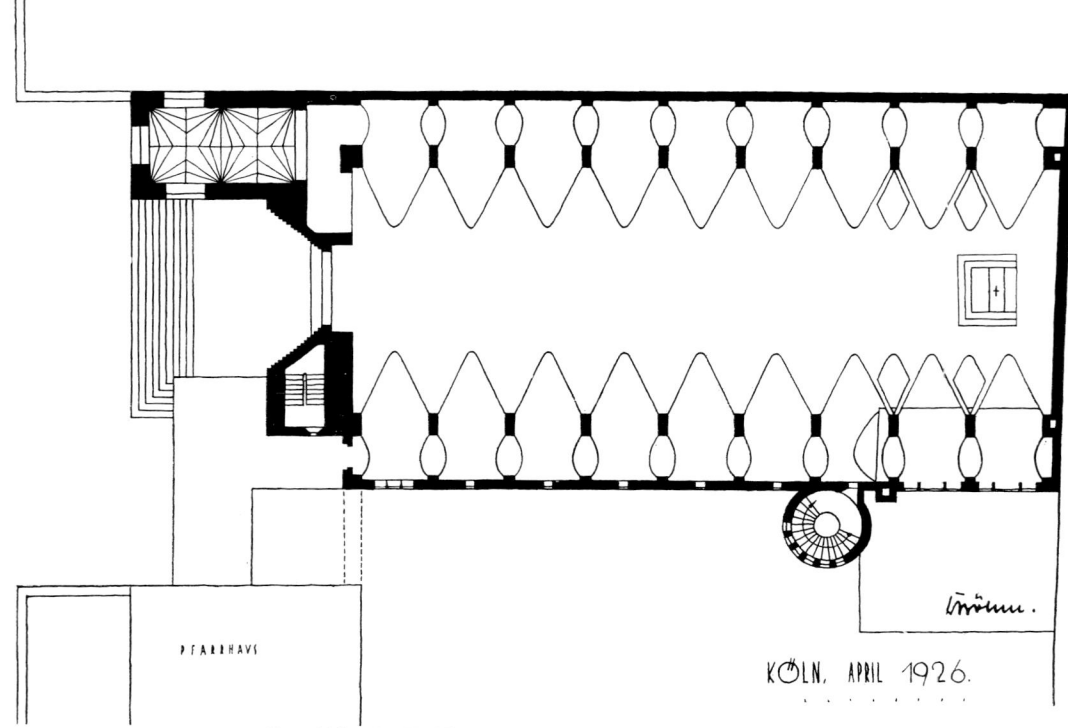

Fig. 316. CATHOLIC CHURCH, BISCHOFSHEIM
Plan.

Fig. 317-318. CATHOLIC CHURCH, BISCHOFSHEIM
Views of Aisles Seen from the Nave.

Fig. 319. CATHOLIC CHURCH, BISCHOFSHEIM
Aisle.
Prof. D. Bohm, Architect.

network of arches covering the vault is unique. The Neu-Ulm church owes its effect to the light-and-shade rhythms produced by cut-in planes; the exterior wall has projections resembling in section the teeth of a saw. These lines in addition to the diagonally placed piers serve to direct all attention to the altar in the apse. In the Resurrection Chapel similar "teeth" are curved as if the whole dome had been twisted (Fig. 331). The brick and stone facings of both churches should have been omitted to stress the monolithic character of the structure.

The parabolic arch is characteristic of Ferro-Concrete which in its absolute freedom to accept any form is well adapted to the ever changing curvature of the parabola. The parabola in turn expresses the monolithic quality of reinforced concrete by merging sides and top in one unbroken

Fig. 320. KRIEGERGEDACHTNIS CHURCH, NEU-ULM, GERMANY

The diagonally placed piers and the zig-zag walls (see Fig. 321-322) aim to direct the eye towards the altar.

PROFESSOR DOMINIKUS BOHM, ARCHITECT.

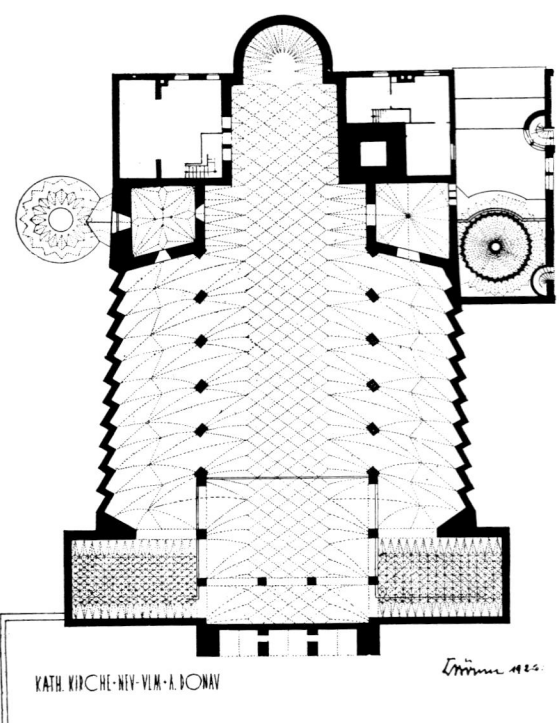

Fig. 321. KRIEGERGEDACHTNIS CHURCH, NEU-ULM
Resurrection Chapel at the left.   Baptistery at the right.
PROF. D. BOHM, ARCHITECT.

Fig. 322. KRIEGERGEDACHTNIS CHURCH, NEU-ULM
Resurrection Chapel in the rear.   Note the "saw-teeth" wall at the right.

Fig. 323. KRIEGERGEDACHTNIS CHURCH, NEU-ULM
Concrete Vaulting over the Narthex.

Fig. 324. KRIEGERGEDACHTNIS CHURCH, NEU-ULM
Narthex Vaulting.

Fig. 325. KRIEGERDACHTNIS CHURCH, NEU-ULM, GERMANY
Aisle.
Note the zig-zag wall on the left and the diagonally placed piers on the right.
PROF. D. BOHM, ARCHITECT.

Fig. 327. BAPTISTERY OF THE KRIEGER-
GEDACHTNIS CHURCH, NEU-ULM
Eye at the crown of the vault seen from below.

Fig. 326. BAPTISTERY OF THE KRIEGERGEDACHTNIS CHURCH,
NEU-ULM

Fig. 329. KRIEGERGEDÄCHTNIS CHURCH, NEU-ULM
Pulpit and side-altar seen from the center of the nave. The Resurrection Chapel is visible behind the rear arch.

Fig. 330. RESURRECTION CHAPEL, NEU-ULM
Crown of vault seen from below.

Fig. 328. KRIEGERGEDÄCHTNIS CHURCH, NEU-ULM
Aisle-Piers seen from the Nave.
PROF. D. BÖHM, ARCHITECT.

Fig. 331. RESURRECTION CHAPEL, KRIEGERGEDACHTNIS CHURCH, NEU-ULM
Prof. Dominikus Bohm, Architect.

HILL AUDITORIUM, UNIVERSITY OF MICHIGAN, ANN ARBOR
The parabola has gained in importance in our Age of Science not only for arches and vaults but also in planning. Halls which have a parabolic plan with the speaker's stand at the focal point have the best acoustics. Hill auditorium is a rotation-paraboloid with curved back.
ALBERT KAHN AND ERNEST WILBY, ARCHITECTS.

curve. Moreover, due to our modern knowledge of statics, the parabolic arch is being increasingly used as the most economical one. These are the main arguments in support of the parabolic arch. Psychological reasons, based on the extension of our limits in time and space which create a new attitude towards the infinite, seem to demand the use of the semi-infinite curve, the parabola, in our age. This concurrence of structural and esthetic reasons indicates that the parabola will more and more become the typical outline for arches and vaults in the Ferro-Concrete Style.

CHAPTER V
THE FERRO-CONCRETE STYLE

As a mountain stream dammed by a fallen tree rises, and rises slowly but surely until sufficient power has been accumulated to sweep away the obstacle, so, works great and small, well done with concrete, will accumulate day by day until the prestige of concrete has developed power enough to sweep away all that is in the way of its natural advancement to its place in architecture and in art as the most satisfying medium which has yet been devised.

J. J. Earley.

Fig. 332. UNIVERSITY OF MUNICH, GERMANY
Hall.
Coffered ferro-concrete vaults.
PROF. G. BESTELMEYER, ARCHITECT.
1909

Fig. 333. FEDERAL GRAIN-BIN, ALTDORF, SWITZERLAND
Mushroom Construction.
MAILLART & CIE, ENGINEERS.

Structural Possibilities.

The advantages of reinforced concrete for construction are so widely recognized that it will suffice to mention a few outstanding features.

Engineer W. W. Clifford points out that for ordinary commercial construction ferro-concrete requires only half as many columns as wood, and is materially less expensive than steel. With heavy loads it is cheaper than wood construction. Steel costs less only in very tall buildings and where live loads are very light and spans long.

Coffered ceilings have a structural justification when produced in reinforced concrete and ought to be utilized by the architect who strives for beauty and sincerity (Fig. 332, 99, 100, 102-105, 119).

Mushroom-construction (Fig. 333-334) has developed some types of columns, with spreading capitals supporting flat ceiling slabs, that promise stylistic development. W. W. Clifford summarizes their advantages in the *Architectural Forum*.[70]

"The girderless floor is unequalled for economy with square bays of from 16 to 30 feet on a side and live loads of 150 pounds per square foot or more. Since the slab rests directly on the spread column heads the ceiling is clear and smooth, which makes for the better distribution of light and heat . . . . By placing the spandrel beam above the slab, as is commonly done, sash may extend to the ceiling which is a very great advantage where a deep room is to be lighted. The thinner floor means less height, floor to floor, for

(70) September 1922, p. 149.

Fig. 333A. ST. ANN'S PARISH CHURCH, WOLLASTON, MASS.

Courtesy "The Western Architect." W. B. Colleary, Architect.

Fig. 334. WAREHOUSE OF THE CANADIAN RAIL AND HARBOUR TERMINALS, TORONTO, CANADA
Mushroom Construction.

the same clear height will often allow 11 stories in a building which would have but 10 with beam framing."

The common practice in Central Europe of placing a sound-insulating three inch thick layer of cinders or burnt building debris on the floor, into which the sleepers which carry the finish-floor are embedded, should be employed for all concrete floors to make them sound-proof.[71] An advantage of concrete buildings is the fact that they are vermin-proof.

As a concrete structure consists of a series of rigid frames bearing the floor, the arrangement of rooms is easily changed as the thin partition-walls can be taken away and others set up in a different position. Furthermore the arrangement of rooms on the upper stories can differ from that of the stories below, as only the columns run thru. Ferro-concrete permits a light type of construction that increases the floor space and proves economical

---
(71) A sketch showing the application of this type of insulation to a typical American wooden joist floor is shown in "Architecture", June 1928, p. 360.

Fig. 335. ATTIC-STAIR, OFFICE BUILDING OF HENSCHEL & SON, KASSEL
CURT VON BROCKE, ARCHITECT.

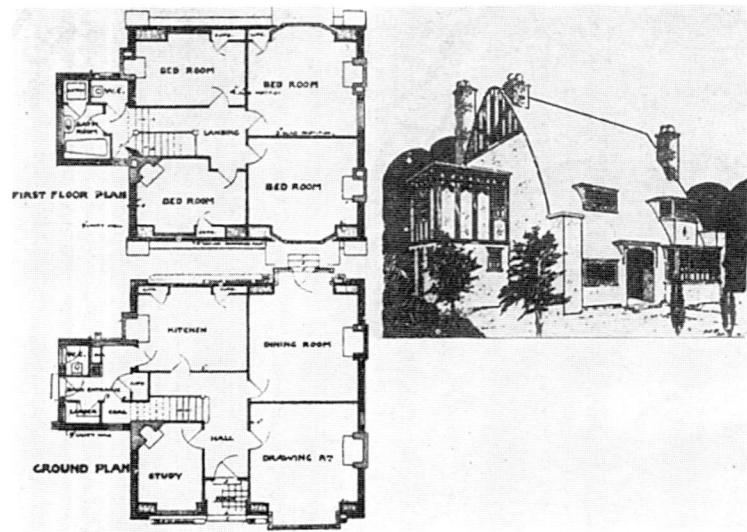

Fig. 336. "CAT-AR SYSTEM" HOUSE, ENGLAND

in reducing the sections of the columns in the lower stories and foundations. The ease with which reinforced concrete is formed into wide overhanging balconies has been mentioned in the first chapter. These assets are common to steel construction too, but ferro-concrete is fireproof and does not require repainting as steel does. Reinforced concrete increases in strength with age, and when properly designed resists earthquakes better than other materials. Ferro-concrete buildings are monolithic and hence, when part of the foundation settles, act like a ship so that a house may actually slant considerably without being damaged. Mr. Twelvetrees reports three buildings near Tunis which had an inclination of 25 degrees and were then restored to a vertical position. In the United States

Fig. 337. PRIMAVERA BUILDING, PARIS EXPOSITION OF DECORATIVE ARTS, 1925
Tiles of different tints are inserted in the concrete dome.
SAUVAGE & WYBE, ARCHITECTS.

Fig. 338. COUNTRY HOUSE "DE SLAH", BLUEMENDAAL, HOLLAND
CHR. BARTEL, ARCHITECT.

engineers have met with similar cases. Considered as to depreciation concrete is unexcelled; it is the most permanent form of construction known.

That the architect must prefer ferro-concrete for stylistic reasons has been well expressed by the Viennese Professor Leopold Bauer:

"Nothing has done more harm to architectural style, unity and character than the universal use of steel beams; thru the ease of their application they replaced all organic types of construction like vaulting, etc. and created a corrupt system of building which, but for far-reaching falsifications would have offered the eye unbearable aspects ... The esthetic difference between massive masonry and slender steel beams could never be reconciled. Ferro-concrete has bridged this gap in a wonderful way by uniting all advantages of steel-construction with the possibility of a really esthetic mode of building. ... There hardly exists a structural problem which it cannot solve with a certain grace, an actually visible static."

Ferro-concrete changes the traditional roof. As the entire construction is fireproof, and the roof-frame has no obstructing struts, the attic is replaced by an additional floor, the

Fig. 339. RESIDENCE IN PASSY, PARIS
RICHARD, ARCHITECT.

227

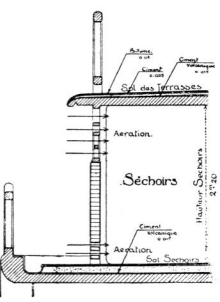
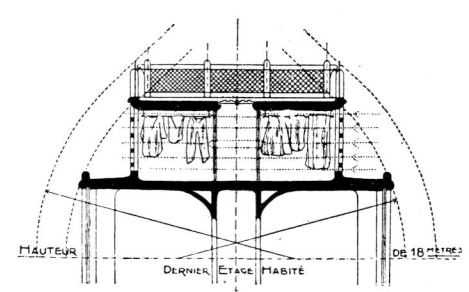

Fig. 340. ATTIC FOR DRYING LAUNDRY COMBINED WITH ROOF-TERRACE
Ferro-concrete construction.
AUG. REY, ARCHITECT.

ceiling of which serves as roof. The reinforced concrete roof is simply a continuation of the wall with a waterproof skin over it. The choice of a porous concrete or the closing of the airpockets between the beams assures insulation against external temperature (Fig. 335). The "Cat-Ar" houses (Fig. 336) outwardly express the unity of the enclosing shell with which ferro-concrete replaces the separate wall and roof of other materials and styles. Transitional types are shown in Fig. 337 and 338, the latter demonstrating a similar idea for brick construction. The opposite extreme, the flat roof, has great possibilities from the architectural viewpoint as the stepped-back building demonstrates (Fig. 339-344, 350, 352, 38, 39). But still more important is its practical value as a roof-garden. Above the dust of the street and open to the sun it is the most healthy spot of the house; pergolas create shady corners for rest, and breezes which

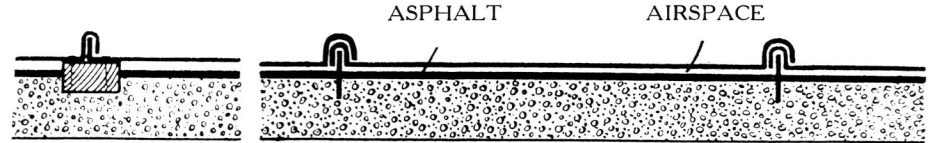
Fig. 341. METAL ROOFING ON CONCRETE SLABS.
(European Examples)

Fig. 342. THE FLAMINGO HOTEL, MIAMI BEACH, FLA.
Cement Stucco Finish.

228

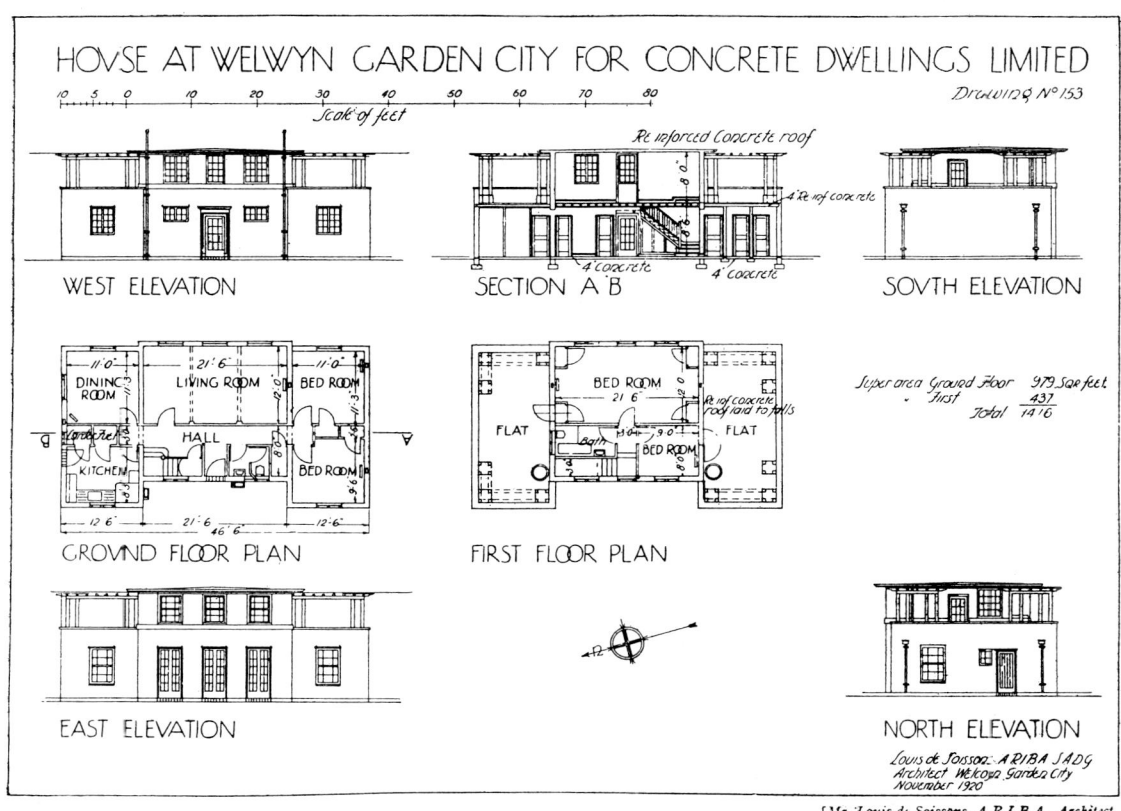

From "Concrete Cottages, Bungalows, Garages", Concrete Publications Ltd., London.

Fig. 343-344. CONCRETE HOUSE AT WELWYN GARDEN CITY, ENGLAND

Louis de Soissons, A. R. I. B. A., Arch.

Fig. 345. CLOTHES FACTORY, PARIS
Avenue Philippe Auguste.
A. & G. PERRET, ARCHITECTS.
1919

are most apt to be felt at this highest plane refresh on a summer evening. Frank Lloyd Wright, Le Corbusier, Mallet-Stevens, and numerous modernists in Germany have recognized that the flat roof is an asset to the house and is valuable as an additional floor.[72] A down-spout located near the center of the roof seems the best way of draining it. In houses having a central heating system water will not be apt to freeze in such a spout, which must have a sieve at the head to prevent leaves from clogging it.

The artistic possibilities of concrete blocks are now being more widely recognized as Fig. 163-174 show. Mr. Allen Wright recommends coating the exposed surface of concrete blocks with waterproofing in the shape of Government whitewash, or one of the several patented cement paints.[73] Waterproofing should be also added to the mortar used for the joints and the latter should not carry across the web. Furring strips and lath can be replaced by corkboard that is secured to the blockwall with either hot pitch or a rich cement mortar; corkboard is nailable and therefore should be used on concrete block walls. A double wall with the inner nonbearing shell three or four inches thick provides a space in which heating-ducts or pipes, and electrical work or insulating material can be placed.

(72) See "Das Werk", Mai 1928, pp. 147-150.
(73) "Architecture", March 1928.

Fig. 346. LA BUTTE-AUX-CAILLES BATHING ESTABLISHMENT, PARIS

Swimming Pool.

Note the glass-block inserts in the vault.

L. BONNIER, ARCHITECT.

Fig. 347. LA BUTTE-AUX-CAILLES BATHING ESTABLISHMENT, PARIS
Swimming Pool.
L. BONNIER, ARCHITECT.

Good and Bad in Ferro-Concrete Design

The many possibilities created by reinforced concrete can also be dangerous because they permit architects to indulge in stunts. In the historical styles the small intercolumniation served as unit for rhythm, but the span between two ferro-concrete columns is too big to act as scale.[74] Ferro-concrete has encouraged some "modernists" to create monstrosities, but as Lewis Mumford writes, "It is the lesser artists and architects who, unable to control and mold the products of the machine, have glorified it in its nakedness." No other architecture exposes the architect's ability as fully as reinforced concrete does; no other material permits such brutality and crudeness. The freaks of a future degenerate period will far surpass those of the Baroque, as ferro-concrete gives the designer more liberty. Even now the shapes given to the new material by some architects remind one of the manners of a parvenu.

To replace the chaos of modernism by a Ferro-Concrete Style, architects would have to agree on what is good and what is bad in design, and herein L. N. Tolstoy's criterion of art could be applied to architecture: The spectator must be infected by the feeling

---
(74) O. C. Hering recommends massing sculptured or incrusted ornamentation at salient points to counteract the impression of leanness and frailty that might be obtained from the great distance between the piers, from the shallow reveals of the openings and from the thin lines of shadow. "Concrete and Stucco Houses", p. 92.

Fig. 348. HOLLYWOOD STORAGE WAREHOUSE, HOLLYWOOD, CAL.
Morgan, Walls & Clements, Architects.

which the artist experienced while creating. Following are some of the sentiments which an architect can express:

1) All experiences common to humanity such as reverence, religion, etc.
2) Sympathy with the struggle and balancing of forces in the building.
3) Sensing the conquering of space by a structure,—the striving of the finials to pierce the sky, of the cornice to project another foot.

The sentiments of the first type can be expressed by sculpture, stained glass, mosaic and tile representations as well as by inscriptions. Concrete-tracery with its black-and-white silhouettes will encourage this telling of a story.

Buildings such as those shown in Fig. 27, 28, and 349 which stress structural facts feature the sentiments of the second type. The semi-

Fig. 349. MUNICIPAL SAVINGS BANK, FREIBURG i. BREISGAU, GERMANY
The inset shows the section of two types of ribs of this vault. (Dimensions are in the metric system).

Fig. 350. ELKS' CLUB, LOS ANGELES
Ferro-concrete with Portland Cement Stucco.
CURLETT & BEEKMAN, ARCHITECTS.

parabola consoles that merge the wireless towers of the Hollywood Terminal Building (Fig. 348) into the main mass satisfy a structural feeling. The arches of the French buildings shown in Fig. 346, 347, tho not strictly parabolic, have the forcefulness of sincere construction.

Some of the tendencies known as "cubism" can be understood as expressing sentiments of the third group. Yet heavy concrete slabs of wide overhang (Fig. 352, 13, 14) conflict with the second type of feeling as they do not show the balancing of forces: the layman believes the slab is about to fall on him and that is hardly an esthetic feeling. Consoles curved according to the moment-curve, a semi-parabola, and carried well down, supporting the overhanging slab would be better structurally and satisfy the man in the street (Fig. 27, 28). Good architecture must tell a tale, be it of the mechanical forces involved, or of the dramatic struggles of mankind.

Architect H. Ranzenberger believes that his Goethaneum (Fig. 354-356) introduces a new style which can only be fully appreciated by adherents of Anthroposophy, a German offshoot of the Theosophist Movement. Insofar as this temple stresses the monolithic character of concrete, has curved planes, and courageous overhangs, and liberates its apertures from the conventional rectangular outline, it demonstrates some essentials of the Ferro-Concrete Style.

Fig. 351. ELKS' CLUB, LOS ANGELES
Detail.

But in spite of these characteristics it is an example of the misuse of the possibilities of liquid stone. Its illogical crudeness should not be blamed on reinforced concrete. A chimney shaped to resemble wreathing flames is a reconciling feature.

The St. Antonius Church (Fig. 357-362) shows that its designers were courageous pioneers and followed the stylistic tenet of using ferro-concrete thruout the building. Its vaulted nave is dignified and sincere; the overhanging, stepped-off organ-balcony which creates deeply recessed portals for the driveway is a unique conception and very typical of concrete. Yet the boxlike tower and the monotonous rectangular grilles of the belfry must be blamed on cubism and similar modern movements, but not on the material employed. A comparative glance at the churches of A.-G. Perret and of Professor D. Bohm will show more typical examples of the evolving Ferro-Concrete Style.

Courtesy *"Literary Digest"*
Fig. 352. CONCRETE HOUSE, RUE NANSOUTY, PARIS
ANDRE LURCAT, ARCHITECT.

Some architects still think that our generation can be satisfied to build in the historical styles and that those who attempt to design in a new way are just seeking publicity. Certain so-called "modern" buildings justify this suspicion. On the other hand it must not be forgotten that shapes and ornaments, just as phrases, when repeated over and over again, utterly

Courtesy *The Architects' Journal (London)*
Fig. 353. HALL WITH CONCRETE STAIRCASE
ANDRE LURCAT, ARCHITECT.

Fig. 354-356. THE GOETHANEUM, ANTHROPOSOPHIC TEMPLE, DORNACH, SWITZERLAND
Under construction 1927.
H. RANZENBERGER, ARCHITECT.

fail to impress. If a building is to make an appeal to the public, it must first gain its attention; to do this the architect is justified in using new expressions for old truths. From this viewpoint the parabolic arch and concrete tracery may be helpful innovations. All depends on the spirit and the artistic standards of the architect. To quote from I. K. Pond:

"The possibilities of texture, the possibilities of color inhering in the product make it a thing through which the sensitive designer can make his feelings flow. So that to have this product made the medium of a wonderfully expressive art all we would seem to need is a wonderfully sensitive designer!"

Fig. 357A. ST. ANTONIUS CHURCH, BASLE
Professor K. Moser, Architect.

Featuring the Monolithic Character of Concrete

The ferro-concrete building is a monolith: column and beam, arch and wall become *one* stone, in which connecting steel rods link the most distant parts together (Fig. 363). Hence separating moldings have no justification. One element of the construction must curve into the other and all angles be chamfered as in the Breslau Market Hall (Fig. 280). Dr. A. Willnow's sketch (Fig. 364) is an original suggestion for stating the same fact by ornament.

E. Mendelsohn's Einstein Tower (Fig. 365-367), called by Frank

Fig. 358. ST. ANTONIUS CHURCH, BASLE
Nave.
PROF. KARL MOSER, G. DOPPLER & SON, ARCHITECTS.

Fig. 357. ST. ANTONIUS CHURCH, BASLE, SWITZERLAND
Tower Seen from Northeast.

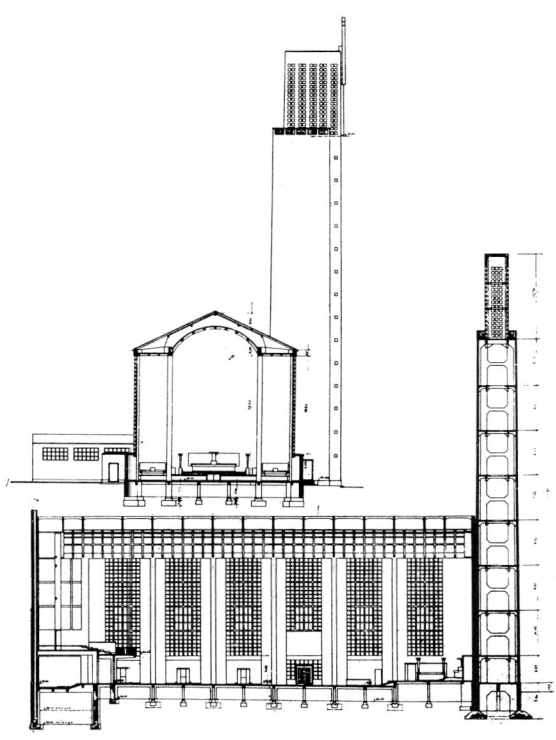

Fig. 359-360. ST. ANTONIUS CHURCH, BASLE
Transverse and Longitudinal Sections.
Prof. K. Moser and G. Doppler & Son, Architects.

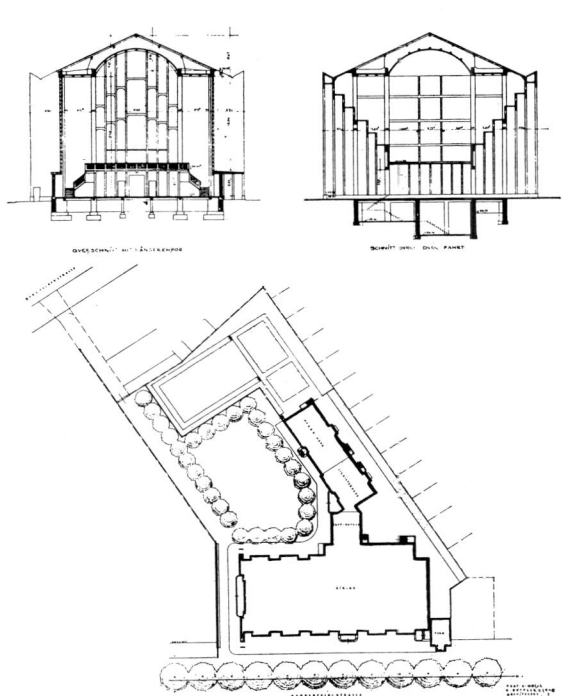

Fig. 361-362. ST. ANTONIUS CHURCH, BASLE
Upper left: *Section with church gallery in the rear.*
Upper right: *Section thru the driveway and organ-loft, (at the extreme left in the plan below).*

Lloyd Wright "a purely plastic structure", is an excellent example of the expression of this monolithic character. Paul Jamot points out that this building does not show the lightness and airiness of reinforced concrete, the most docile and resisting of all materials.

The absence of all "seams", the possibility of unlimited expanse, and the plasticity of ferro-concrete permit slight swellings as well as the gradual rising and fading away of forms. Thus the pierced coffers of the Karlsruhe Station vault are echoed in the shallow coffers at the springline, which grow deeper and deeper as the vault ascends (Fig. 368). The smokestack of the Blatt & Haselbach mills is in its lower part a cylinder with trefoil-swellings which gradually vanish into a purely circular cylinder at the top (Fig. 369).

Reinforced concrete permits modelling on a large scale; its monolithic character is also shown in the splays and cut-in surfaces which give such an effective play of light and shade (Fig. 370, 371, 320-331, 349).

One of the main effects of a concrete structure lies in its unity. A building that is to contribute to the Ferro-Concrete Style should be made completely of reinforced concrete and glass, except for the usual fittings of other materials. As Frank Lloyd Wright states, "But the more the building tends toward a mono-material building—the more nearly will perfect style reward an organic plan and ease of execution economize results. The more logical will the whole become."

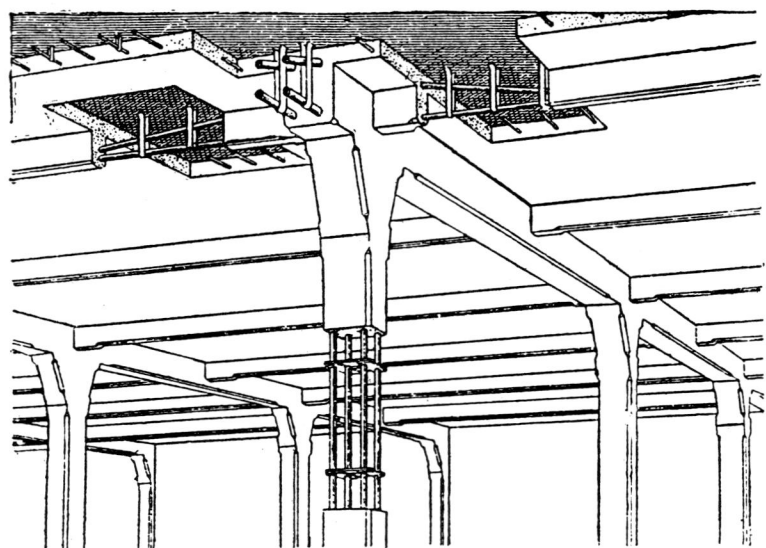

Fig. 363. FERRO-CONCRETE CONSTRUCTION

R. B. Williamson words this axiom as follows:

"In such masterpieces as the Parthenon and St. Paul's Cathedral is a definite relation between structure and ornamentation. Construction and design were both conceived in the same material—cut stone. ... This leads to an important truth—one might call it an axiom regarding architecture. In all real art, structure and design form an inseparable unity. The close relationship between them makes it impossible successfully to divorce one from the other. And the most effective ornamentation must be wrought in the material used for the construction. ... For the most part the buildings of the future will be constructed and designed in concrete. The unity between structure and ornamentation will be effected through the use of this material, which will serve both the engineer and the designer."[75]

"I can recall ancient structures of equal importance", says John J. Earley, "of similar and dissimilar character built in the years just before or after the beginning of the Christian Era, whose foundations were one material, whose superstructures were another and whose architectural finishes were many; also I can recall but cannot understand modern structures built by these ancient methods .... Recently I visited one of our large modern cathedrals, now under construction. It is being built by the same methods and with the same materials as was the Colosseum at Rome, but I noticed with more than ordinary interest that skewed arches and such forms were re-

---

(75) "Concrete", April 1927.

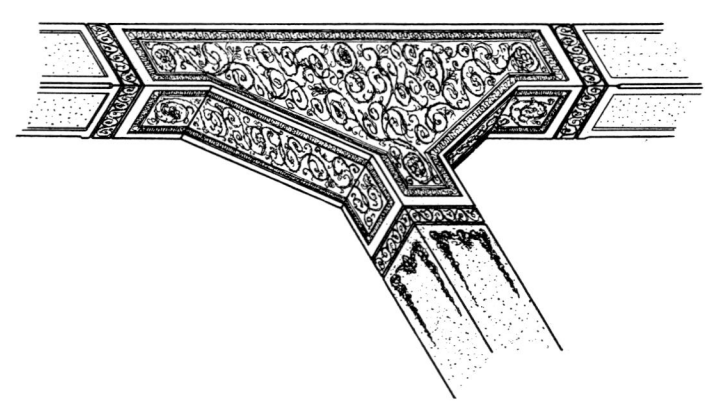

Fig. 364. DESIGN FOR A FERRO-CONCRETE BEAM-COLUMN INTERSECTION
Dr. A. Willnow, Architect.

Fig. 365. EINSTEIN TOWER, POTSDAM, GERMANY
Viewed from the West.
ERICH MENDELSOHN, ARCHITECT.

Fig. 366-367. EINSTEIN TOWER, POTSDAM, GERMANY
Details.
Erich Mendelsohn, Architect.

Fig. 368. RAILROAD STATION, KARLSRUHE, GERMANY
The dado consists of polished concrete (see Fig. 121)
STURZMACHER, ARCHITECT.

inforced concrete. Certainly if concrete is valuable in these places it is valuable elsewhere and, had the designer been of that mind, it might have been used generally with benefit to all."

Mr. Woollett correctly claims:

"Properly handled a building built entirely of concrete should cost less all things considered than one built by the ordinary method which is to build a frame of concrete and iron, to which we tie, bolt, screw and hang the various finish materials such as iron, terra cotta, brick, furring, plaster, marble, etc. A giant hand could pass into a modern theatre constructed in this manner and sweep away all of the so-called finish, leaving only the bare concrete structure. In short we build two buildings: one a complete structural entity, the other an elaborate architectural and decorative falsehood. . . . All the architecture of past ages has been more or less real in the sense that the construction and the decoration are part and parcel of each other and has expressed in one operation, the idea of building and the idea of beauty. We have yet to solve this definite problem, i. e. to express in beautiful forms the type of construction thrust upon us by a utilitarian age."

Featuring the Reinforcement.

If the tale of the stresses in a ferroconcrete building are to be made intelligible, the existence and location of reinforicng bars should be indicated by "veins" in the surface as described in chapter II. Furthermore the sectional-area percentage of steel used as reinforcement in any one building should be kept constant as far as structural demands permit, so that the bigger load would always be supported by the wider pier and the

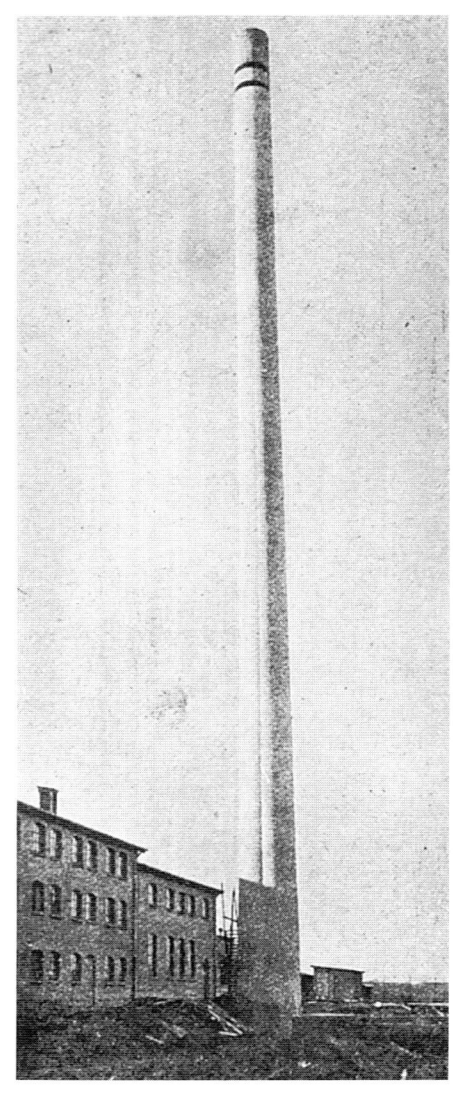

Fig. 369. SMOKESTACK, BLATT & HASSELBACH MILLS, LYNGBY, DENMARK
(48 meters high).

Fig. 370. OFFICE BUILDING, HENSCHEL & SON, KASSEL
Entrance.
Tooled concrete surface.

Fig. 371. LUMBER-STOREHOUSE, LOCOMOTIVE WORKS, HENSCHEL & SON, KASSEL, GERMANY
Curt von Brocke, Architect.

higher beam. Lastly, reinforcing should be reduced to the allowable minimum as it is the stone element in concrete which gives it a monumental and architectural quality. Thus in a monumental type of building the arch is preferable to the strongly reinforced beam; the paraboloid is preferable to the cupola having a heavily reinforced tension-ring to obviate buttressing. As Cass Gilbert states, "A building.... should not only be strong, but it should *look* strong. A very small percentage of increase in the cost will accomplish this."[76]

In 1908 the Hungarian Medgyaszay presented the following valuable suggestions to the VIIIth International Architect's Congress: "When we in-

---
(76) "The Architectural Forum", Sept. 1923, p. 86.

Fig. 372. THEATRE, VESZPREM, HUNGARY
Note the tracery in the consoles.

Fig. 373. VESZPREM THEATRE, HUNGARY
Terrace and Pergola.
(See Fig. 195-197).
Prof. I. Medgyaszay, Architect.

sert steel bars in the parts bearing tension, the compression parts will be about twenty times more massive than the tension parts. This is a specific ferro-concrete characteristic as expressed for instance in the ribbed ceiling construction. . . . When a painter or a sculptor depicts an action, he chooses the most characteristic moment, and that is one of the last moments. A column of elastic material will swell under pressure just before failing: the convex outline is the one most characteristic of pressure . . . The entasis of the Greek column is a small but essential finesse; such finesses are the products of living, feeling souls, of the artist. An elastic bar in tension will become more and more slender, and, if of homogeneous material, most of all in the center: both outlines become concave and now the shape of the bar is most characteristic of tension. This is the second principle for the artistic treatment of ferro-concrete. This principle demands that all supports and the upper chords of trusses be formed slightly convex; i. e. they should receive an entasis. The lower chords and tiebeams should have a "negative entasis", that is, they should be slightly concave. These main outlines should approximate straight lines so as not to deny the rigidity of the material. The brackets (Fig. 372)

Fig. 374. HOUSE ON ORLAY-UTCAI, BUDAPEST
Note the concrete tracery of the balconies.
PROF. ISTVAN MEDGYASZAY, ARCHITECT.
1910

Fig. 375. MODEL OF TOMB-CHURCH IN NOGRADSZAKAL, HUNGARY
The walls are of concrete blocks. All ornamental parts as well as the tower are of monolithic concrete.
PROF. I. MEDGYASZAY, ARCHITECT.
1911

carry the concrete slab and balustrade: the upper chord bears tension, the lower one compression. Applying the above principle, the upper chord tapers towards its center and the lower one is swelled. Of the balcony-beams (Fig. 373-374) which stiffen the columns the top ones in some cases carry tension and the outline is therefore continually concave. In the same instance the bottom side would be in compression and this is here expressed by small notches as if the material was about to fold due to pressure."[77]

Medgyaszay was pleased to notice that the lower members of the trusses in the Breslau Markethall (Fig. 280), being tension-members, have a "negative entasis". He strives to contrast the bearing elements, such as columns, beams, ribs and consoles, with the filling planes which serve to separate the spaces. These planes he treated either as uniform slabs, or pierced them when they acted as balustrade or window (Fig. 196-199, 375).

The Ferro-Concrete Style.

The Ferro-Concrete Style will be a new type of Gothic if the contentions of this book prove to be true. Not the historic Gothic forms, but its unconquerable spirit will be resurrected thru ferro-concrete. The pointed arch will be replaced by its sister,—the parabolic arch; the stone tracery of windows by the reinforced concrete tracery of entire walls; the paint that covered Gothic masonry by the colored aggregates of surface-layers. As Gothic was based on the unity of material—stone—so the new

---

(77) "Reports of the VIII. International Architects' Congress, 1908".

248

Fig. 376. UNITED STATES ARMY SUPPLY BASE, BROOKLYN, N. Y.
"*There is something very fine about a great gray mass of building, all one color, all one tone, yet modified by the sunlight or shadow to pearly gray of wonderful delicacy.*"—Cass Gilbert, Architect.

Fig. 377. U. S. ARMY SUPPLY BASE, BROOKLYN, N. Y.
"*It is the big simplicity of the thing that counts, and if there may be projections for necessary fire towers or elevator shafts, or other salients, if there may be low roof structures for tank houses or machinery, and if the glass surfaces are kept in scale, there may be silhouette, and light and shade.*"—Cass Gilbert, Architect.

style will be characterized by ferro-concrete prevailing from foundation-pile to roof-balustrade, from chimney-flue to wall-tracery. The Ferro-Concrete Style will be Gothic as John Ruskin defined that term:

"For in one point of view Gothic is not only the best, but the *only rational* architecture, as being that which can fit itself most easily to all services, vulgar or noble. Undefined in its slope of roof, height of shaft, breadth of arch, or disposition of ground plan, it can shrink into a turret, expand into a hall, coil into a staircase, or spring into a spire, with undegraded grace and unexhausted energy; and whenever it finds occasion for change in its form or purpose, it submits to it without the slightest sense of loss either to its unity or majesty,—subtle and flexible like a fiery serpent, but ever attentive to the voice of the charmer."[78]

(78) "The Stones of Venice", II-37.

Courtesy *Journal of the American Institute of Architects*
Fig. 378a-b. BUILDINGS IN OPA-LOCKA, FLA.
The stuccoed concrete domes are blue or soft browns, the colors graduating upward to white or cream.
BERNHARDT E. MULLER, ARCHITECT. 1925

Courtesy *The Architectural Record*
Fig. 379a. SWIMMING POOL, OVERBROOK, PA.

Courtesy *The Architectural Record*
Fig. 379b. SWIMMING POOL, OVERBROOK, PA.
J. I. BRIGHT & STERNFELD, ARCHITECTS.

Fig. 380. CENTENARY HALL, BRESLAU, GERMANY
Note in the section shown in the inset that the upper part of the dome rests on roller-bearings.
STADTBAURAT BERG, ARCHITECT.
1913

In his book on the brothers Perret, Paul Jamot stresses the similarity of Concrete Architecture with Gothic: "It is necessary that concrete govern, that everywhere it let its power and special virtues be realized. Obviously it favors the vertical more than any other line.[79] Gothic architecture, which from other motives has lead to the same predilection shows us what beauty, wealth and variety verticalism can bestow."

More than any other material reinforced concrete can express the superiority of man's mind. For many possibilities mean many opportunities where the human will can choose and decide.

If a Ferro-Concrete Style is to develop, architects must concentrate on the use of this material. The many possibilities of concrete require a thoro investigation by the artist who wants to unite them harmoniously. In no other material is a craftsman's knowledge so essential for success as with liquid stone, where the qualities and quantities of the component parts must be determined correctly to achieve good results.

The many "modernisms", historical styles, and techniques offered to our architects have a bewildering effect and prevent excellence in any one

---
(79) p. 75. I believe the parabolic curve more typical.

252

Fig. 381. CENTENARY HALL, BRESLAU

Fig. 382. CENTENARY HALL, BRESLAU
Detail of Auditorium.

field. John Ruskin, who could not foresee the development of ferro-concrete, recommended that architects agree on one historical style so as to end the present chaos. He points out how the architecture of a nation is only great when it is as common as its language; that in our present state of doubt and ignorance we can hardly realize how being limited to one style would cause an awakening of thoughts, a growing feeling of ease and power, and of the true kind of freedom. Liberated from the unrest and torture of choice, the architect would be able to enter into the final secrets of the accepted style and would find his understanding broadened, his practical knowledge secure and applicable, and his imagination enlivened and incited as that of a child in a walled garden.[80] By limiting ourselves to the Ferro-Concrete Style the world would become our "walled garden". Gothic flourished

(80) "The Seven Lamps of Architecture".

Fig. 383. CENTENARY HALL, BRESLAU
Max Berg, Architect.

thruout Europe but the new style will develop thruout the whole world; for reinforced concrete is now already used on every continent—in Bombay as well as in Stockholm, in the Argentine as well as in Russia. The Ferro-Concrete Style has developed with telegraph and railroad, with motion picture and radio; it is a child of the age that created the League of Nations, and witnessed transoceanic flights.[81] Unity is its essence, and concrete, replacing the piling of stone upon stone, brick upon brick by the creation of one unit—the complete building—, was given us in time to express this yearning of our generation. The Ferro-Concrete building with its millions of pebbles and sand-specks molded into a monolith is a symbol of united humanity. Climatic differences and historic backgrounds will prevent monotony, so that we must picture the Ferro-Concrete Style as depicting "unity in diversity."

The preceding chapters might create the erroneous impression that the development of a Reinforced Concrete Style is a question of molding methods, surface treatment and calculation. On the contrary, as Mecenseffy, Riegel and others state, the purpose of a building and the method of its construction are not the most important style-creating agencies as some claim. The conception of the artist, the indwelling spirit directing his imagination are the decisive factors, as Frank Lloyd Wright expresses so clearly:

"No matter how technically faithful his logic may have been to his scale and

---

(81) The general adoption of Esperanto as auxiliary international language would promote correspondence and interchange of literature and thus help the development of Ferro-Concrete Architecture—the International Style.

materials and method—over and above all that, living in the atmosphere created by the orchestration of those matters, hovers the indefinable quality of style. ... Usually you hear music as you work."

As Dean Emil Lorch of the University of Michigan pointed out, style is the expression of an era, of the impulses controlling man in a given period.

Of all materials, ferro-concrete offers the greatest wealth of possibilities to the designer. But these possibilities remain unrealized until sensed by a living soul. Hence the Ferro-Concrete Style will not come into its own until architects the world over have awakened to the Spirit of the New Age.

SUBJECT INDEX

Aberrations..........................1
Aggregates....................17, 125
    ", colored (see "Surface Layers")................66, 69
    ", lightweight................23
Airpockets........................120
Airspaces (double-wall)........44, 178, 207
Atterbury System...................49
Auto-centring...................35-43
Aztec Style..................120, 122
Balconies.........................23
Balustrades......................248
Baroque..................24, 50, 232
Beams.......................35, 119
    ", false......................89
Blocks (concrete)...47, 59, 123, 167, 204, 230
Brick-molds......................42
Bridges.........................189
    ", Parabolic..............192, 198
Bubblestone......................23
Cantilevers......................26
    " —vaults....................26
Casting (of monument)............124
Cat-Ar-House............204, 209, 228
Cement-film..............55, 59, 71, 72
Cement-Gun..........28, 44, 47, 87, 130
    " ", Centring................44
Centrifugal force methods..........47
Ceramics.........................71
Chiseled concrete................127
"Cinderella"..................50, 53
Coffered Ceilings................223
Colored aggregate.........71, 73, 78, 79, 81
Colored concrete...........61, 66, 102, 165
Columns. 26, 36, 47, 119, 174, 223, 232, 247, 248
    " —brackets..................188
    ", tapering.................187, 210
Compressed air technique...........44
Crudeness of design..32, 34, 52, 53, 232, 236
Cubism.....................24, 235, 236
Curved shapes. 43, 49, 97, 142, 147, 188, 235, 239
Dangers in design................232
Domes...23, 26, 36, 39, 42, 46, 119, 135, 209
Economy.....................34, 35, 223

Efflorescence.....................54
Entasis (positive and negative)..........247
Esperanto.......................255
Esthetics of Potentialities...............
    ....17, 24, 177, 210 (footnote)
Etching concrete.................129
Exposing aggregate.............55, 59
Facing Layers, see "Surface-layers"
Failure of foundation............226
Flexibility of arrangement.......225
Floor-systems........15, 26, 35, 39, 42, 223
Form-marks............35, 54, 55, 59, 81
Forms, removing of...............127
Galleries........................26
Glasin..........................102
Glass Blocks...........157–159, 162, 212
Glass inserts.................100, 101
Glazing.....................102, 105
Glue molds............119, 120, 122, 134
Good design.........232, 233, 235, 246
Gothic........3, 15, 24, 32, 128, 138, 162,
   177, 178, 197, 198, 207, 248, 250, 252, 254
Gravity system....................44
Haircracks....................54, 59
Hall-type....................187, 189
Hardening, retarded..............128
Intarsia.........................97
Kerament........................105
League of Nations................255
"Liquid Stone"............49, 131, 252
Luxfer-grille....................156
Masts...........................47
Metal forms......................35
Metallized Concrete..............105
Metric System...................156
Mineral colors................62, 65
Mixing aggregates.................66
    " colors......................65
Mixtures (proportions)...............
    ....43, 69, 87, 99, 122, 23, 128
Modelling concrete....47, 130, 134, 237, 241
Modernism...................232, 252
Moldings..................112, 127, 239
Monolithic Character...239, 241, see "Unity"
Mushroom construction............223

257

SUBJECT INDEX—Continued

Newness of Ferro-Concrete Style. 26, 177, 180
" " " " ", American authorities................3, 5, 8, 11
Newness of Ferro-Concrete Style, European Authorities................11, 13
Ornaments........................133
Outrigger Construction.............26
Painting..........................88
" , examples..........91–94, 99, 100
" , paints........90, 99, 100, 123, 230
" , preparation of surface......89, 90
" , protective coats..............91
Panels.......................108, 119
Parabolic arch........186–221, 238, 248
" " , economy...............192
" " , esthetic value..........195
" " , examples.......198, 202, 204
" " , German estheticians. 196, 197
" " , gracefulness........189, 204
" " , Greek Estheticians.......196
" " , headroom...............209
" " , historical precedents. 195, 204
" " , infinity........195, 196, 198
" " , psychological effect......195
" " , structural advantages....192
Parabolic arcades................207, 208
" churches...................210
" consoles...................235
" details....................204
" doors......................207
" plan.......................221
" windows...................207
Paraboloid........................246
Passivity of concrete...............8
Patents..........................32
Permanent molds................36–42
Permeability......................69
Pigments.....................61, 62, 65
Placement......................43, 49
Plaster molds........119, 120, 125, 127
Plasticity.........29, 30, 32, 34, 106, 241
Polishing.............95, 101, 106, 177
Possibilities.........17–51, 23, 252, 256
Precast concrete........36, 47, 49, 66, 108
Projections.........23, 24, 34, 49, 150, 233

Prejudices..............15, 17, 30, 51, 178
Qualitative variations.............21–23
Quantitative variations............17–21
Raincy, Notre Dame du...........166–177
Reinforcing........42, 43, 49, 180, 245–247
" in sculpture..............108
" veins.....................95
Reliefs..........................120
Reveals..........................108
Rigid bent } 187, 209
" frame }
Roman concrete....................3
Roofgarden.....................228, 230
Roofs.................26, 204, 227, 230
Sandblasts.......................59
Sand-molds......................120
Sawdust molds..................35, 41
Saw-teeth wall...................214
Selfbearing walls.................27
Self-centring......30, 42, 43, 46, 47, 159, 212
Self-supporting reinforcement........42, 46
Setbacks.........................27
Scraffito......................87, 88
Scrubbing........................59
Sculpture...............106–130, 166
" , economy of...........106, 128
" , possibilities.............129
Sound-insulation..................225
Space-Art...................24, 150, 151
Splays...........................241
Spraying.........................95
Stairs, surface-layer...............23
" , spiral....................162
Stauss-bricklets..................39
Steel............22, 50, see "Reinforcing"
" , area.....................17, 245
" , kinds of..................21
Structural possibilities...........223
" advantages................220
Stucco......................82, 84, 87
Surface........................52, 53
Surface-layers........23, 66, 69, 127, 248
Textile-block-slab...............123, 151
Textures.........................53
Tile.......................97, 100, 122

SUBJECT INDEX—Continued

Tooling..............................59
Tracery 32, 41, 123, 238, 248, 133–185
    "   casting...............134, 137, 156
    "   clarity.......................155
    "   details of execution............150
    "   Egyptian reliefs...............148
    "   figural...............138, 177, 233
    "   historical scenes...............154
    "   office buildings................151
    "   pictorial, see "Figural"
    "   psychological reasons......133, 138
    "   sculpture.....................154
    "   transitional types.............134
    "   ventilation...........134, 147, 156
    "   windows. 134, 139, 151, 154, 155, 156
Trusses..............................248
Turning..............................47

Unity of Ferro-Concrete........23, 35, 50, 55, 137, 187, 214, 239, 241, 242, 245, 243, 255
Variability of concrete............17, 20–23
Vaults............32, 34, 159, 178, 192, 236
    ", parabolic.........204, 207, 210, 241
Veneers......................42, 52, 214
V-frame.............................188
Washing concrete......................87
Water-cement-ratio....................20
Waterproofing..................123, 230
Water, removing from fresh c....27, 69, 125
Weathering.......................52, 53
White concrete........................71
Wood-centring...................49, 112
Wood-centring Style. 27, 30, 32, 34, 43, 124, 188
Workability......................20, 34

INDEX TO ARCHITECTS, ENGINEERS, SCULPTORS, AND AUTHORS

	ILLUSTRATIONS Page	TEXT Page
Abrams, Duff A.		20
Ackerman, F. L.		5, 30
Aichinger & Schmidt	157	
Allen, Glenn	81–83	99, 100
Allison & Allison	63, 64, 69, 96	8, 50, 87, 88, 93, 129, 131
Atterbury, Grosvenor	45, 54, 146	49, 135
Austin, John C.	61, 62, 130	8, 192
Badovici Jean		11
Bakewell & Brown	34, 40, 41	
Bates & Wigglesworth	148	
Baudot de	42	42, 134
Bauer Leopold		227
Berg Max	252–255	
Bestelmeyer, G.	222	
Bohm, Dominikus	211–220	39, 198, 210, 236
Bonatz, P.	105	
Bonnier, L.	173, 231, 232	
Bourgeois, L.	145	
Bright, J. I., & Sternfeld	250, 251	
Brillant, Maurice		175
Brocke, Curt von	18, 188, 189, 225, 246	
Brostrom, E. O.	200	
Brower, T. A.	1	
Brunner Associates (Gehron, Ross & Alley)	71–73	91
Clifford, Walter W.		65, 89, 112, 187, 223
Coate, R. E.	80	
Colleary, W. B.	224	
Cram & Ferguson	198	
Crowel, J., Jr	206	
Curlett & Beekman	235, 236	
Davison, R. C.		47, 61
Dulfer, M.	158	
Durant		196
Earley, John J.	38, 56–59.	11, 20, 52, 61, 71, 81, 95, 106, 124, 125, 222, 242
Eberson & Weaver	97, 154	
Edison, T. A.		35
Emperger, F. von	193	
Fechner, Theodor		197, 212
Fischer, Th.	160	
Flagg, Ernest		17
Fletcher, Sir Bannister		148, 207
Foster, W. D.		167, 168
Freyssinet	198	

	ILLUSTRATIONS Page	TEXT Page
Garnsey, J. E., & Parsons	68	
Gilbert, Cass	249	5, 30, 59, 246
Goethe, W.		145
Goodhue, B. G.	65, 67, 68	26, 150
Greenley, Howard	197	
Hall, R. E., & Co.	19–36	
Harmon, A. L.		138
Harnack, A. von		196
Harvey, A. E.	95, 177	
Havlik, R. F.		59
Hegel		196
Hellmann & Littman	162	
Hering, O. C.		8, 52, 65, 81, 188
Hildebrandt, Ad. von	37	
Hill, S. Woods	25–28	3, 5, 26, 35
Hoffman, Josef	133, 199	
Hogarth		198
Holabird & Roche	85, 86	
Honore, Paul	171	
Hooper, J.		100
Horta, M. V.	172	
Howard, John G.	132	
Jamot, Paul		13, 175, 177, 178, 180, 241, 252
Janssen		108
Johnson, Nathan C.	71	92
Johnson, W. T., & Snyder, R. W.	38	
Kahn, Albert	79, 161, 221	5, 150
Kiehnel & Elliot	166–170	154
Korner, C.		192
Kreis, Wilhelm, Dr.	18, 186, 199	198
Kuster, Dr.	191	192
Le, Coeur F.	174	
Le, Corbusier		230
Lakeman, A.		35
Lindberg, J.	204	
Lorch, Emil		256
Lurcat, Andre		11, 139
Lux		196
Maillart & Cie	223	
Mallet-Stevens		230
Margotin-Thierot, L.	16, 17	
Marrost & Droz	140–144	
Marshall & Fox	70	
Mason, J. B.		8, 88, 89
Mecenseffy, E. von		13, 39, 255
Medgyaszay, I.	134–138, 247, 248	13, 134, 246–248
Mendelsohn, Erich	186, 187, 243, 244	239

	ILLUSTRATIONS Page	TEXT Page
Meyer & Holler	46–48	
Moczelany	3	
Monier, Joseph		3, 23
Morgan, Walls & Clements	12, 101-104, 106-111, 233	122
Moser, Karl	239–241	236
Muller, B. E.	250	
Mumford, Lewis		232
Newell, F. V.	28, 148	
Nordenkampf, B. von	175	159
Onderdonk, F. S.	148	
Palanti, Mario	11	
Perret, A and G.	84, 175–185, 230	3, 13, 36, 55, 162, 177, 178, 180, 204, 236
Pfeiffer, E.	37	
Pinand, J. H.	202–204	
Pite, Beresford		5
Plato		196
Plotin		196
Plumet, Ch.	15, 16	192
Poelzig	14	
Pond, I. K.		28, 238
Popper-Lynkeus		195
Portaluppi, Piero	159	
Price & McLanahan	2	
Pusch, G.	201	
Pythagoras		196
Rank Bros.	192	
Ranzenberger, H.	238	235
Rey, A.	228	17
Richard	227	
Riegel		255
Riepert, P. H.		13
Riphahn, W.	56	
Rodrigues, D.	51, 131	
Rother	190	
Ruskin, John		95, 139, 149, 155, 195, 250, 254
Rutgers, G. J.	208	
Sarrebezolles	124–129	128
Sauvage & Wybe	226	
Schaeffer, P.	158, 161	
Schuhmacher, F.	53	
Sinclair, Upton		149
Soissons, L. de	229	
Stacy-Judd, R. B.	98, 100	
Sturgis, Russell		30
Sturzmacher	245	

	ILLUSTRATIONS	TEXT
	Page	Page
Taft, Lorado	122	124
Taut, Bruno	205	
Thunnissen	207, 208	
Tolstoy, L. N.		149, 232
Tournon, Paul	124	
Trost & Trost	29–32	
Tuttle, Bloodgood	171	
Twelvetrees		54, 91, 226
Van, Laren J.	208, 209, 210	
Wahlman, L. I.	200	198
Wahl & Roder	50	
Walsh, H. V.	39, 84	
Weithaler, A.	88	102
Wielemans, von	40, 56, 93, 94	36, 112, 134
Wilby, Ernest	79, 161, 221	150
Williamson, R. B.	155	242
Willnow, A.	242	13, 239
Winslow, C. M.	65, 67	
Woollett, W. L.	74, 77	11, 93–95, 125, 126, 245
Woringer		197
Wright, Allen		230
Wright, Frank Lloyd	8, 9, 10, 117–120, 163–65	8, 32, 36, 112, 117–120, 123, 124, 133, 154, 163–165, 173, 174, 230, 241, 255
Wright, Lloyd	112–116	
Zaccagna, G. L.	147	135

BIBLIOGRAPHY

History: W. W. Clifford,	"Concrete Construction", "The Architectural Forum", May 1922, pp. 177, 178.

1905 William Millar,	"Plastering" (Chap. XVIII & XIX), B. T. Batsford, London.
1909 ————	"Concrete Houses and Cottages", 2 volumes, Atlas P. Cement Co., New York.
1910 Ralph C. Davison,	"Concrete Pottery and Garden Furniture", Munn & Co., New York.
1911 E. V. Mecenseffy, Professor,	"Die Kunstlerische Gestaltung der Eisenbetonbauten.— I. Erganzungsband des Handbuches fur Eisenbetonbau", Wilhelm Ernst & S., Berlin.
1912 Oswald C. Hering,	"Concrete and Stucco Houses", McBride, Nast & Co., New York.
1914 Dr. P. H. Riepert,	"Die Architektur im Eisenbetonbau", Cementverlag, Charlottenburg.
1914 Dr. A. Willnow,	"Bisherige Ausbildungen des Eisenbetons im Bau und auf Bauverwandten Gebieten", Wilhelm Ernst & S., Berlin.
1917 ————	"The Manufacture of Standardized Houses", Standardized Housing Corp., New York C.
1918 Albert Lakeman, M. S. A.,	"Concrete Cottages, Bungalows and Garages", Concrete Publications Ltd., London W. C. 2.
1920 Harvey Whipple,	"Concrete Houses and how they were built", "Concrete"—Publishing Co., Chicago.
1924 John J. Earley,	"Architectural Concrete", American Concrete Institute, Detroit.
	"Substance, Form & Color through Concrete", Atlas Portland Cement Co., New York.
1927 T. P. Bennett, F. R. I. B. A.,	"Architectural Design in Concrete", Oxford University Press, New York.
1927 Paul Jamot,	"A.-G. Perret et L'Architecture du Beton Arme", G. Vanoest, Paris.

Publications of the Portland Cement Association, Chicago:

    J. J. Earley, "The Concrete of the Architect and Sculptor".

    H. C. Mercer, "The Decoration of Concrete with Colored Clays", (Bulletin No. 10).

"Portland Cement Stucco" (1926).

"Concrete in Architecture" (1927).

— — — — — — — — — —

Bibliography of *Surface Treatment* compiled by W. W. Clifford:*

In a general way in "The Handbook of Building Construction" and in "Concrete Engineers Handbook" by Hool & Johnson.

Proceedings of the American Concrete Institute:
- Vol. XVI. J. J. Earley and J. C. Pearson, New Developments in Surface Treated Concrete and Stucco".
- Vol. XVII. John W. Lowell, "Coloring Concrete". Committee report on treatment of Concrete Surfaces. "Shrinkage of Portland Cement Mortars and its Importance in Stucco Construction."

U. S. Bureau of Standards, Technologic Paper No. 70 (Results of a winter's exposure on 56 stucco panels).

Atlas Portland Cement Co., "Cast Stone".

Universal Portland Cement Co., "Concrete Surfaces".

— — — — — — — — — —

A bibliography of Reinforced Concrete *Construction* is contained in Th. Crane "Concrete Building Construction", J. Wiley & Sons, New York, 1927.

(*) "The Architectural Forum", Feb. 1922, p. 70.